# INFORMATION SECURITY
# MANAGEMENT SYSTEMS

*A Novel Framework and Software*
*as a Tool for Compliance with*
*Information Security Standards*

# INFORMATION SECURITY MANAGEMENT SYSTEMS

## A Novel Framework and Software as a Tool for Compliance with Information Security Standards

Heru Susanto, PhD
Mohammad Nabil Almunawar, PhD

APPLE
ACADEMIC
PRESS

Apple Academic Press Inc.
3333 Mistwell Crescent
Oakville, ON L6L 0A2 Canada

Apple Academic Press Inc.
9 Spinnaker Way
Waretown, NJ 08758 USA

© 2018 by Apple Academic Press, Inc.
First issued in paperback 2021
Exclusive worldwide distribution by CRC Press, a member of Taylor & Francis Group
No claim to original U.S. Government works

ISBN 13: 978-1-77-463652-7 (pbk)
ISBN 13: 978-1-77-188577-5 (hbk)

**Library and Archives Canada Cataloguing in Publication**

Susanto, Heru, 1965-, author
Information security management systems : a novel framework and software as a tool for compliance with information security standards / Heru Susanto, PhD, Mohammad Nabil Almunawar, PhD.

Includes bibliographical references and index.
Issued in print and electronic formats.
ISBN 978-1-77188-577-5 (hardcover).--ISBN 978-1-315-23235-5 (PDF)

1. Management information systems--Security measures.
2. Industries--Security measures--Management. 3. Risk assessment.
I. Almunawar, Mohammad Nabil, author II. Title.

| HD61.5.S87 2017 | 658.4'78 | C2017-905895-9 | C2017-905896-7 |

CIP data on file with US Library of Congress

Apple Academic Press also publishes its books in a variety of electronic formats. Some content that appears in print may not be available in electronic format. For information about Apple Academic Press products, visit our website at **www.appleacademicpress.com** and the CRC Press website at **www.crcpress.com**

# CONTENTS

# ABOUT THE AUTHORS

**Heru Susanto, PhD**
*Head and Researcher, Computational Science & IT Governance Research Group, Indonesian Institute of Sciences; Honorary Professor and Visiting Scholar at the Department of Information Management, College of Management and Hospitality, Tunghai University, Taiwan*

Heru Susanto, PhD, is currently the head and a researcher of the Computational Science & IT Governance Research Group at the Indonesian Institute of Sciences. He is also an Honorary Professor and Visiting Scholar at the Department of Information Management, College of Management and Hospitality, Tunghai University, Taichung, Taiwan. Dr. Heru has experience as an IT professional and as web division head at IT Strategic Management at Indomobil Group Corporation. He has worked as the Prince Muqrin Chair for Information Security Technologies at King Saud University in Riyadh, Saudi Arabia. He received a BSc in Computer Science from Bogor Agricultural University, an MBA in Marketing Management from the School of Business and Management Indonesia, an MSc in Information System from King Saud University, and a PhD in Information Security System from the University of Brunei and King Saud University. His research interests are in the areas of information security, IT governance, computational sciences, business process re-engineering, and e-marketing.

 **Mohammad Nabil Almunawar, PhD**
*Senior Lecturer and Dean, School of Business
and Economics, University of Brunei
Darussalam (UBD), Brunei*

Mohammad Nabil Almunawar, PhD, is currently a senior lecturer and the Dean of the
School of Business and Economics, University
of Brunei Darussalam (UBD), Brunei Darus-
salam. Dr. Almunawar has published more than 60 papers in refereed journals, book chapters, and presentations at international conferences. He has
more than 25 years of teaching experience in the area of computer and
information systems. His overall research interests include applications
of IT in management, electronic business/commerce, health informatics,
information security, and cloud computing. He is also interested in object-
oriented technology, databases and multimedia retrieval.

Dr. Almunawar received his bachelor degree in 1983 from Bogor Agri-
cultural University, Indonesia; his master's degree (MSc in Computer Sci-
ence) from the Department of Computer Science, University of Western
Ontario, London, Canada, in 1991, and a PhD from the University of New
South Wales (School of Computer Science and Engineering, UNSW),
Australia, in 1998.

# LIST OF ABBREVIATIONS

| | |
|---|---|
| 5S2IS | five stages to information security |
| 8FPs | eight fundamental parameters |
| 9STAF | nine state of the art framework |
| ADODB | ActiveX Data Object DataBase |
| BAU | business as usual |
| BoD | Board of Directors |
| BoM | Board of Managers |
| BS | British Standard |
| CIA | Confidentiality Integrity Authority |
| CMM | capability maturity model |
| CMMI | capability maturity model integration |
| CNSS | Committee on National Security Systems |
| COBIT | control objectives for information and related technology |
| COM | component object model |
| COSO | Committee of Sponsoring Organizations |
| DCOM | distributed component object model |
| DDoS | distributed denial of service attacks |
| DMZ | demilitarized zone |
| ECs | essential controls |
| ENISA | European Network and Information Security Agency |
| FGD | focus group discussion |
| FGIS | The Framework for the Governance of Information Security |
| GISPF | The Government Information Security Policy Framework |
| GUI | graphical user interface |
| ICM | implementation checklist method |
| ICT | Information and Communication Technology |
| IEC | International Electronic Commission |
| IEEE | Institute of Electrical and Electronics Engineers |
| IP | internet protocol |
| IPR | intellectual property right |

| IRM | information risk management |
|---|---|
| IS | information systems |
| ISA | information security awareness |
| ISACA | Information Systems Audit and Control Association |
| ISBS | Information Security Breaches Survey |
| ISF | integrated solution framework |
| ISM | Integrated Solution Modeling Software |
| ISMS | Information Security Management System |
| ISO | International Standard Organization |
| ISP | internet service provider |
| ITG | Information Technology Governance |
| ITGA | Information Technology Governance Institute |
| ITIL | Information Technology Infrastructure Library |
| ITMO | Information Technology Manager and Officer |
| ITSCM | Information Technology Service Continuity Management |
| ITSM | Information Technology Services Management |
| MISA | Multimedia Information Security Architecture |
| NIST | National Institute of Standard and Technology |
| OCX | object linking and embedding control extension |
| OLE | object linking and embedding |
| OPM3 | organizational project management maturity model |
| P-CMM | people capability maturity model |
| PCIDSS | Payment Card Industry Data Security Standard |
| PDCA | Plan Do Check Action |
| PMBOK | project management body of knowledge |
| PMC | Prince Muqrin Chair for Information Security Technologies |
| PMMM | project management maturity model |
| PRINCE2 | Projects in Controlled Environments – Version 2 |
| PWC | Price Waterhouse Cooper Consultants |
| QGIA | Queensland Governance of Information Assurance |
| QGISPF | Queensland Government Information Security Policy Framework |
| REM | release and evaluation methodology |
| RISC | readiness and information security capabilities |
| RM | research methodology |

| | |
|---|---|
| RMA | release management approach |
| SAM | security assessment management |
| SDA | spiral development approach |
| SDLC | Software Development Life-Cycle |
| SEPG | Software Engineering Process Group |
| SIEM | security information and event management |
| SIM | security information management |
| SMM | security monitoring management |
| SOA | service oriented architecture |
| SoA | statement of applicability |
| SP | software performance |
| SPP | software performance parameter |
| SQ | software quality |
| SQL | structure query language |
| SSAD | Security Systems Analyst and Developer |
| STOPE | Stakeholder Technology Organization People Environment |
| TCP | transmission control protocol |
| TOGAF | The Open Group Architecture Framework |
| URS | user requirement specification |
| VB | Visual Basic |
| VOOP | visual object oriented programming |
| WFA | waterfall approach |
| WSP-SM | waterfall software process-spiral model development |

# LIST OF TABLES

# LIST OF FIGURES

# PREFACE

Information security contributes to the success of organizations, as it gives a solid foundation to increase both efficiency and productivity. Many business organizations realize that compliance with the information security standards will affect their business prospects. Securing information resources from unauthorized access is extremely important. Information security needs to be managed in a proper and systematic manner as information security is quite complex. One of the effective ways to manage information security is to comply with an information security management standard. There are a number of security standards around; however, ISO 27001 is the most widely accepted one. Therefore, it is important for an organization to implement ISO 27001 to address information security issues comprehensively. Unfortunately, the existing ISO 27001 compliance methods are complex, time consuming and expensive. A new method, preferably supported by an automated tool, will be much welcomed.

One of the key components for the success of information security certification is by using a framework. This framework acts as a tool to understand the process and technical aspects. Unfortunately, existing frameworks do not provide fixed and practical models for RISC (Readiness and Information Security Capabilities) investigation, which is investigation conducted to find out an organization's readiness and information security capabilities regarding ISO 27001.

This study proposes a novel framework called the Integrated Solution for Information Security Framework (ISF). ISF was developed to tackle issues that are not properly addressed by existing security frameworks for RISC investigation and provides an easy and practical model for information system security according to ISO 27001. Based on ISF, a semi-automated tool is developed to assess the readiness of an organization to comply with ISO 27001 and subsequently use the tool to assess the potential threats, strengths and weaknesses for efficient and effective implementation of ISO 27001. This tool is called Integration Solution Modeling Software (ISM), which is based on ISF, to assist organizations in measuring the level of compliance of their information systems with ISO 27001. The software consists of two major modules: e-assessment to assess the level of compliance with ISO 27001; and e-monitoring to monitor suspected activities that may lead to security breaches.

ISM provides the ability to enhance organizations beyond usual practices and offers a suitable approach to accelerate compliance processes for information security. ISM brings a possibility to enhance organizations by enabling them to prepare for the processes of security standardization by conducting self-assessment. A new approach in ISM helps organizations improve their compliance processes by reducing time, conducting RISC self-assessment, handling SoA preparation, monitoring networks, and suspect detection monitoring.

To see the effectiveness of ISF and ISM, we conducted a comprehensive ISM testing and evaluation. The result is very promising as ISM is highly regarded and accepted as a useful tool to help companies systematically plan to acquire ISO 27001 certification. User responses towards the performance, quality, features, reliability, and usability (called by eight fundamental parameters – 8FPs) are high. Overall score according to 8FPs is 2.70 out of 4, which means close to "highly recommended." ISM performs RISC investigation within 12 hours, which is much better then implementation a checklist method (ICM – the currently existing method to measure RISC level in the organization) approaches that require approximately 12 months for the investigations. This means that our framework is effective, and certainly its implementation is useful for organization to assess their compliance with ISO 27001 and to set a clear strategy to obtain ISO 27001 certification with confidence.

# COMMENTARIES

Comments on published papers from academicians, editors, and professionals are delineated below. Those papers are part of this work.

*"I recommend this work on this topic. The authors have lots of knowledge, and the topic is important. Security in IT usually is access controlled and consists of authentication and authorization."*
**—Prof. Dr. Günter Müller**
*Institute of Computer Sciences and Social Studies,*
*Department Telematics,*
*University of Freiburg, Germany*


*"We consider the content and your approach very valuable. We came to the conclusion that the level of knowledge you have lead to a good chance to overcome the hurdles of the next steps. We are confident with your work will have the chance to become a really appreciated contribution to the scientific and practical IS community."*
**—Prof. Dr. Martin Bichler**
*Department of Informatics,*
*Technische Universität München, Germany*

# CHAPTER 1

# INTRODUCTION

## CONTENTS

### 1.1  STUDY OVERVIEW

We are living in the information age, where information and knowledge are becoming increasingly important and no-one denies that information and knowledge are important assets that need to be protected from unauthorized users such as hackers, phishers, social engineers, viruses, and worms that threaten organizations on all sides, through intranet, extranet, and the Internet. The rapid advancement of information and communications technology (ICT) and the growing dependence of organizations on ICT continuously intensify concern on information security (Von Solms, 2001). Although, most ICT systems are designed to have a considerable amount of strength in order to sustain and assist organizations in protecting information from security threats, they are not completely immune from the threats (Furnell, 2005). Organizations pay increasing attention to information protection as the impact of information security breaches today have a more tangible effect (Dlamini et al., 2009; Furnell et al., 2006; Furnell & Karweni, 1999).

Cherdantseva et al. (2011) and Pipkin (2000) looked at information security from the business standpoint and argued that information security needs to be considered as a business enabler and become an integral part of business processes. Von Solms (2005), Tsiakis & Stephanides (2005), and Pipkins (2000) stated that information security may help to raise trust in an organization from customers and it should be understood that security of information brings many advantages to business (e.g., improved efficiency due to the exploitation of new technologies and increased trust from partners and customers). Saint-Germain (2005) argued that an important driver for information security management system adoption is to demonstrate to partners that the company has identified and measured their security risks, implemented a security policy and controls that will mitigate these risks, also to protect business assets in order to support the achievement of business objectives (Boehmer, 2008; Dhillon, 2007; Furnell et al., 2006; Saleh et al., 2007a, 2007b).

Cherdantseva & Hilton (2013), and Sherwood et al. (2005) adopted a multidimensional and enterprise-wide approach to information security and proposed to include a wider scope of information security covering various aspects of business such as marketing and customer service. Information security is no longer considered purely from a technical perspective, but also from a managerial, system architect's and designer's points of view and it could enable businesses to increase competitiveness (Sherwood et al., 2005), economic investment (Anderson, 2001; Gordon & Loeb, 2002; Tsiakis & Stephanides, 2005), products or services to world markets transparently and in compliance with prevalent standards, such as ISO 27001 and ISO 17799 (Theoharidou et al., 2005).

It is clear that information security needs to be managed properly as related issues are quite complex. Several information security management system standards were developed to assist organizations in managing the security of their information system assets. It is important to adopt an information security management system (ISMS) standard to manage the security of organization's information assets effectively. In contrast, Standish Group (2013) stated that many ICT projects in the US, including ISMS standardizing and ISO 27001 compliance in major organizations, faced difficulties, with many having reported failure and only around one in eight (13%) ICT projects attempting to standardize information security were successful. Othman et al. (2011), and Fomin et al. (2008) stated that technical barriers, the project owner's 'absence of understanding processes, technical aspects, lack of internal ownership and neglect of certain aspects were major problems that caused the delay for these ISMS and ISO 27001 projects. An organization may face challenges in implementing an ISMS standard without proper planning, and any obstacles could create roadblocks for effective information security adoption (Kosutic, 2010, 2013), such as:

- *Financial issues*. At first sight, it may seem that paperwork should not cost too much, until the stakeholder realizes that they have to pay for consultants, buy literature, train employees, invest in software and equipment.
- *Human resources issues*. The expertise dedicated to implement ISMS is unavailable.
- *Participation issues*. An ISMS adoption project may be seen as solely the initiative of an ICT department rather than the engagement of the entire organization.
- *Communications issues*. Lack of proper communication at all levels of the organization during the ISMS certification process.
- *Technical issues*. Translation of the technical terms and concepts of a chosen ISMS standard is required. Essential controls dealing with the standard are very technical and will not be readily understood by the board of management as decision maker, making it difficult to be implemented by an organization. Therefore, those terms need to be refined, otherwise the controls will tend to be somewhat disorganized and disjointed.
- *Selection and adoption issues*. Difficulty in selecting a suitable ISMS standard for related organizations. There are several standards for IT Governance which lead to information security such as PRINCE2, OPM3, CMMI, P-CMM, PMMM, ISO 27001, BS7799, PCIDSS, COSO, SOA, ITIL and COBIT. It indicates that an organization has to choose the best standard that is suitable for their business processes and also well-recognized by their partners, clients, customers, and vendors.

As mentioned above, several challenges arise when implementing the standard. One of the key components to understanding the process and technical aspects is by using a framework to support ISMS and ISO 27001 projects. Although the development of ICT security frameworks has gained momentum in recent years, more work on approaches to security framework are still needed, as the current frameworks do not provide measurements to assess the readiness level of organizations to adopt an ISMS standard (Calder & Watkins, 2012; Calder et al., 2010; Fomin et al., 2008; Potter & Beard, 2010).

To fill the gap, this study proposes a novel approach and develops a system that can measure the closeness of an organization's information security status with an ISMS standard (a compliance level). This framework is designed in such a way to derive an integrated solution to overcome the organization's technical barriers and difficulties in understanding, investigating, and complying with an ISMS standard (ISO 27001). This framework, called Integrated Solution Framework (ISF), helps organizations map the assessment issues, controls, and clauses of ISO 27001 to its related domain and acts as a measurement tool for assessing the information security compliance level of organizations toward ISO 27001.

ISF consists of 6 main components identified as domains, namely: organization (domain 1), stakeholders (domain 2), tools & technology (domain 3), policy (domain 4), culture (domain 5), knowledge (domain 6). Those are associated with the critical components within an organization that relates to information security circumstances, and further ISO 27001 compliance stages. The explanations for each domain are expanded in Chapter 4: Proposed Framework.

Based on ISF, the assessment and monitoring software was developed, called Integrated Solution Modeling (ISM). This software measures the RISC[1] level of an organization towards ISO 27001, analyzes security events in real time, and collects, stores, and reports for regulatory compliance. The software has two main functions:

1. Security assessment management (SAM/e-Assessment). Log management and compliance reporting. SAM provides the collection, reporting and analysis of assessment data that will show the

---

[1] Readiness and Information Security Capabilities

strength and weakness points and increase priority on low achievement points to support regulatory compliance.

2. Security monitoring management (SMM/e-Monitoring). SMM
monitors real-time activity, firewall and network management to
provide monitoring and identify potential security breaches. ISM
collects network activity data in real time so that immediate analysis can be done.

To make sure the effectiveness of the framework (ISF) and its implementation (ISM) in assisting organizations, we conducted comprehensive
testing on the reliability, usability, and performance in respondent organizations in the field of telecommunications, banking & finance, airlines,
and ICT-security consultancy. The results of the testing and evaluation
were further analyzed using software performance parameters (SPP) and
release and evaluation management (REM) to find out the software performance, features and quality, to obtain a RISC measurement (Bakry, 2003a,
2003b; Herbsleb et al., 1997). There are eight defined parameters to measure the performance and features of the framework and software (Bakry
2001, 2004; Gan, 2006; McCall et al., 1977a, 1977b) as follows: (1) How
ISM functions in information security self-assessment; (2) The benefits
brought by ISM in helping organizations understand ISMS standard (ISO
27001) controls; (3) How ISM can be used to find out information security
terms and concepts; (4) ISM features; (5) ISM graphical user interface and
user friendliness; (6) Precision of the analysis produced by ISM; (7) Final
result precision produced by ISM; (8) ISM performance.

## 1.2 THE SCOPE OF THE PROBLEM AND MOTIVATIONS

There are many important questions associated with organizations and
security standards in relation to security awareness and compliance. This
study proposes a framework as a solution for the technical aspects of the
research questions:

1. What are the main barriers in implementing ISMS within an
organization?
2. What are the differences between existing state-of-the-art frameworks and solutions to formal and quantitative investigation of
RISC parameters, and what are their weaknesses?

3. How significant the proposed framework will reduce the learning and preparation time as the organization enhances itself for ISO 27001 compliance?

4. What are the main advantages for an organization in self-assessing using ISM to obtain the RISC measurement regarding ISO 27001 certification?

The motivation of this study is to improve the overall ability of organizations to participate, forecast, and actively assess their information security circumstances. Enhancement is one of key indicators for improving readiness and capabilities of information security. The organization's enhancements provide users the ability to conduct self-investigation and real-time monitoring of network activities. The current RISC investigation tool uses the ICM[2] approach. In some case studies, organizations spent approximately 12 months to conduct RISC investigation. On the other hand, Kosutic (2012) stated that for RISC investigation of compliance processes, organizations commonly take between 3–36 months.

Many organizations experience difficulty in implementing and complying with an ISMS standard, including obstacles faced when measuring the readiness level of an organizational implementation, document preparation as well as the various scenarios and information security strategies to deal with (Susanto et al., 2011a; Siponen & Willison, 2009). An organization may face internal and external challenges in implementing an ISMS standard. Without proper planning, the following obstacles could create a barricade for effective information security implementation (Furnell, 2005; Kosutic, 2012; Susanto et al., 2011a, 2012b, Von Solm, 2001):

1. Expertise and employment of it may be beyond an organization's capability.

2. Difficulty in selecting existing information security standards, for instance in choosing out of PRINCE2, OPM3, CMMI,P-CMM, PMMM, ISO 27001, BS7799, PCIDSS, COSO, SOA, ITIL or COBIT. Each standard plays its own role and position in ISMS, such as (1) information security associated with the project management and IT governance, (2) information security

---

[2] Implementation Checklist Method.

related to business transactions and smart cards, and (3) overall information security management system as the main focus of the standard.

3. Compliance with an ISMS standard such as ISO 27001 requires all employees to embrace new security controls introduced by the standard.

## 1.3 RESEARCH POSITIONING

This study is related to information security management system standards, risk management associated with information security and information security awareness within an organization. The details are explained in the following subsection.

### 1.3.1 INFORMATION SECURITY MANAGEMENT SYSTEM

An ISMS is a set of policies concerned with information management and ICT risks. The governing principle behind an ISMS is that an organization should design, implement and maintain a coherent set of policies, processes and systems to manage risks to its information assets, thus ensuring acceptable levels of information security risk. As with management processes, an ISMS must remain effective and efficient in the long-term, adapting to changes in the internal organization and external environment (Kelleher & Hall, 2005). The establishment, maintenance, and continuous update of the ISMS provide a strong indication that an organization is using a systematic approach for the identification, assessment, and management of information security risks and breaches.

The chief objective of ISMS is to implement the appropriate measurements in order to eliminate or minimize the impact that various security related threats and vulnerabilities might have on an organization. ISMS will enable implementation of desirable characteristics of the services offered by the organization (i.e., availability of services, preservation of data confidentiality and integrity, etc.). However, the implementation of an ISMS entails the following steps: definition of security policy, definition of ISMS scope, risk assessment, risk management, selection of appropriate controls, and statement of applicability (Calder & Watkins, 2010; Potter & Beard, 2012). To be effective, efficient, and influential towards an organization's business processes, ISMS implementation must follow scenarios such as:

- It must have the continuous, unshakeable and visible support and commitment of the organization's top management;
- It must be an integral part of the overall management of the organization related to and reflecting the organization's approach to risk management, the control objectives and controls and the degree of assurance required;
- It must have security objectives and activities based on business objectives and requirements and led by business management;
- It must fully comply with the organization's philosophy and mindset by providing a system that instead of preventing people from doing what they are employed to do, it will enable them to do it in control and demonstrate their fulfilled accountabilities;
- It must be based on continuous training and awareness of staff and avoid the use of disciplinary measures;
- It must be a never ending process.

There are several ISMS standards that can be used as benchmarks for information system security. An organization can choose one of these standards to comply with. The big five of ISMS standards (Susanto et al., 2011a) are ISO 27001, BS 7799, PCIDSS, ITIL and COBIT. Susanto et al. (2011b) stated that ISO 27001 is the ISMS standard most widely used globally. ISO 27001 specifies requirements for the establishment, implementation, monitoring and review, maintenance and improvement of a management system – an overall management and control framework – for managing an organization's information security risks.

Moreover, ISO 27001 consists of protection against the following aspects: *Confidentiality* ensuring that information can only be accessed by an authorized person and ensure confidentiality of data sent, received and stored; *Integrity* ensuring that data is not altered without the permission of authorized parties, to maintain the accuracy and integrity of information; *Availability* guarantees that data will be available when needed ensure that legitimate users can use the information and related devices.

### 1.3.2 MANAGING RISK ASSOCIATED WITH INFORMATION SECURITY

Risk Management is a recurrent activity that deals with the analysis, planning, implementation, control and monitoring of implemented measures and enforced security policies (Blakley et al., 2001). It is the process of implementing and maintaining appropriate management controls including policies, procedures and practices to reduce the effects of risk to an acceptable level. The principles of risk management can be directed both to limit adverse outcomes and to achieve desired objectives. Risk management regulates risks toward information and knowledge assets from any internal-external disclosure and unauthorized access, use, disclosure, disruption, modification, perusal, inspection, recording or destruction within an organization. Managing risk associated with information assets is called Information Risk Management (Humphreys et al., 1998).

Moreover, information risk management[3] adapts the generic process of risk management and applies it to the integrity, availability and confidentiality of information assets and the information environment. Information risk management should be incorporated into all decisions in day-to-day operations. Information risk management deals with methodologies and incorporates the typical analysis, assessment, audit, monitoring, and management processes. The details of each stage are as follows (Blakley, 2001; Kelleher & Hall, 2005):

1. **Analysis** examines a given situation, checking for obvious deficits according to professional experience or even common sense. The examination can be structured and repeatable. An information security penetration test and vulnerability scan is an analysis whose purpose is to identify whether the perimeter is vulnerable, identifies flaws, and determines if such a flaw really poses a problem for the organization.
2. **Assessment** identifies a problem and describes how much of a problem it is. A related term in ICT security is vulnerability assessment. As an extension of a vulnerability scan, a vulnerability assessment sets the results of a scan into the context of the organization and

[3] Managing risk associated with information assets is called information risk management. It consolidates property values, claims, policies and exposure of information and management reporting capabilities (Humphreys et al., 1998).

assigns an urgency level. In general, an assessment uses a structured approach, is repeatable, and describes the level of a problem.

3. **Audit** compares a given situation with some sort of standardized situation; an external standard (for instances, a law, or an industry standard) or an internal one (e.g., a policy document). The results of an audit explain how much reality deviates from an expected or required situation.

4. **Monitoring** is an operational activity which introduces the notion of time, as the process of monitoring is real-time and continuous. Proper monitoring requires an established approach to be able to show trends and activities consistently and efficiently.

5. **Management** is a strategic activity. It involves understanding the situation (analysis), determining the extent of the problem (assessment), standardizing the examination (audit), and continuing these activities over time (monitoring). Moreover, it adds the components of remediation, initiating and tracking changes, also includes the necessary communication within the organization.

### 1.3.3  INFORMATION SECURITY AWARENESS

Information security awareness (ISA) is the knowledge and attitude members of an organization possess regarding the protection of the physical, especially information, assets of an organization. According to the European Network and Information Security Agency (ENISA, 2012), ISA is awareness of the risks and available safeguards as the first line of defense for the security of information systems and networks. The focus of security awareness should be to achieve a long-term shift in the attitude of employees towards security, promoting a cultural and behavioral change within an organization. Security policies should be viewed as key enablers and an integral part of a business, not as a series of rules restricting the efficient working of business processes.

Being security-aware means acknowledging that there is the potential for some people to deliberately or accidentally steal, damage, or misuse the data that is stored within a company's computer systems and throughout its organization. Therefore, it would be prudent to support the assets of the institution (information, physical, and personal) by trying to stop that from happening. These following issues especially show the importance of ISA (Kosutic, 2012; Peltier, 2005a, 2005b):

1. The nature of sensitive material and physical assets employees may come in contact with, such as trade secrets, privacy concerns and government classified information.
2. Employee and contractor responsibilities in handling sensitive information, including review of employee nondisclosure agreements.
3. Requirements for proper handling of sensitive material in physical form, including marking, transmission, storage and destruction.
4. Proper methods for protecting sensitive information on ICT systems, including password policy and use of authentication.
5. Other computer security concerns, including malware, phishing, social engineering, etc.
6. Workplace security, including building access, wearing of security badges, reporting of incidents, forbidden articles, etc.
7. Consequences of failure to properly protect information, including potential loss of employment, economic consequences to the firm, damage to individuals whose private records are divulged, and possible civil and criminal penalties.

Information security breaches within organizations were reported by Information Security Breaches Survey (ISBS) (Potter & Beard, 2012), which stated that 'incidents caused by staff' was experienced by 82% of the sampled large organizations (Figure 1.1). No industry sector appears immune from these incidents. Telecommunications, utilities and technology companies appear to have the most reliable systems. The public sector, travel, leisure and entertainment companies are most likely to have security problems. Moreover, it was found that the average security incident within local business organizations occurred once a month, while large or international organizations would expect an incident to occur once a week (Potter & Beard, 2012).

Nowadays, to face with ISA issues, most organizations have allocated more of their budget towards security than in the previous year (2008–2011). On average, organizations spend 8% of their IT budget on information security, and those that suffered a very serious breach were found to

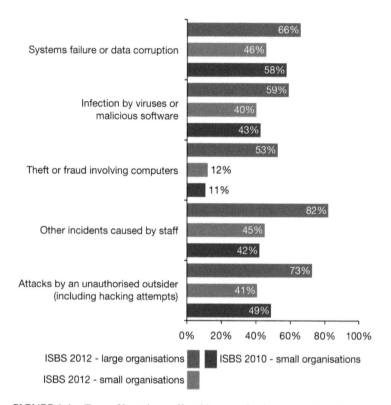

**FIGURE 1.1**   Type of breaches suffered by organizations (ISBS) (Potter & Beard, 2012).

have spent on average 6.5% of their IT budget on security (Potter & Beard, 2012).

As mentioned, ISA is the behavior of employees regarding protection of information assets, such as customer information and customer transactions, therefore having influence on customer trust and customer loyalty. Kottler (2002) and stated, it is obvious that business organizations are dependent on their loyal customers for business sustainability. Customer loyalty is all about attracting the right customers, winning their trust and providing convenience, getting them to buy, buy often, buy in higher quantities, and bring even more customers (Kotler, 2002). ISA implementation should be viewed as one of the corporate efforts, serving the following functions: (1) to improve corporate selling point to customers (Kottler, 1969, 2002); (2) corporate imaging and branding. Corporate branding is an economic-management and social event as well as a strategy through which customers' demands and providers' supplies are balanced (Dwyer et al., 1987); (3) to win the competitive edge within the related business area (Morrison et al., 2003); (4) as one of the marketing tools (Figure 1.2) (Kottler, 2002); (5) to increase corporate profitability (Brown et al., 2000); and (6) to increase customer trust, leading them to become loyal customers stemming from amity and customer satisfaction, sustaining the interdependency between producer and customer (Baker et al., 1996; Brown et al., 2000).

## 1.4 RESEARCH METHOD

This research was performed through literature review, analysis, refinement of ISMS standards, proposed framework (ISF) and implementation of ISF as a software application (ISM). There were several stages conducted. The first stage was knowledge discovery and building knowledge as the first phase of the research, conducted through literature reviews on related work, comparative studies and refinement. The second stage was the construction of a new framework (ISF). The third stage was creating software architecture, constructing variables, assessment of formulae and

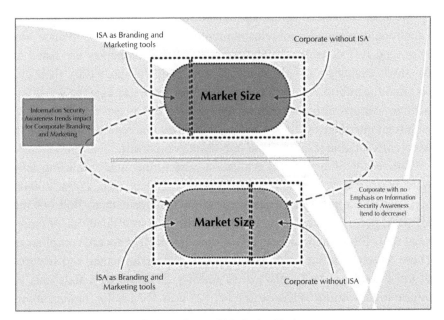

**FIGURE 1.2**    ISA Impact for Branding and Marketing Tools.

software development. The last stage was comprehensive ISM evaluation; this includes testing on reliability, usability, and performance of ISM within the context of an organization.

We conducted testing on a variety of sizes of organizations; small organizations (up to 100 employees), medium sized organizations (101–250 employees) and large organizations (more than 250 employees) (Potter & Beard, 2010) as users of ISF-ISM to find out their preferences and tendencies toward ISM. The companies have businesses in the fields of telecommunications, banking and finance, airlines, and ICT consultants. These organizations were grouped in three categories:

1.  Group I: ISO 27001 holders. Companies that recently received or were certified by ISO 27001 in the period of 2010–2012.
2.  Group II: ISO 27001 ready. Companies currently pursuing ISO 27001 compliance, whether they were in the documents preparation stage, scenario development stage or risk management analysis stage.
3.  Group III: ISO 27001 consultants. Companies in this group are ICT consultants in the security area, particularly information security and standards.

We used a selected sampling method, in which the respondents were intentionally selected from telecommunications, banking and finance, airlines, and ICT consultants. The majority of the companies are listed in the stock exchange and the companies are well recognized by their clients and the public. As listed companies, they have strategies to win competitive markets in the respective industries and they are very concerned with retaining their by clients and customers by maintaining their trust, in which information security is an important component.

The results of testing and evaluation were further analyzed using software performance parameters (SPP) (Bakry 2003a; Gan, 2006; 2003b; McCall, 1977) and release and evaluation management to find out the ISF-ISM performance, features and reliability and its efficacy as measurement tools for an organization's RISC level in ISMS standard compliance. There are eight defined parameters to measure performance and features of the framework and software, as follows: (1) how ISM functions as information security self-assessment? (2) how ISM helps organizations understand ISMS standard (ISO 27001) controls? (3) how ISM can be used to understand information security standard terms and concepts? (4) ISM features; (5) ISM graphical user interface and user friendliness; (6) analysis precision produced by ISM; (7) final result precision produced by ISM; and (8) ISM performance (Bakry, 2003a, 2003b; Gan, 2006; Von Solms, 2001).

A detailed discussion on the methodology of the study is provided in Chapter 3 of this book.

## 1.5  OUTCOME AND CONTRIBUTIONS

One of our research's contributions was observes the barriers facing implementation of an ISMS standard within an organization and identifying the cause of increased numbers and costs of information security breaches that are rising fast. The gaps in existing information security adoption clearly demonstrates the need for the proposed novel approach (ISF) to further appropriate information security awareness, risk management associated with information security, and ISMS compliance (further discussed in Chapter 4: Proposed Framework).

The major contribution of our research is the framework (ISF) and a new measurement approach. This enabled the binding of organizational security policies and standards to the governance and compliance requirements. This contribution changes the landscape of information security standard adoption to a more structured approach and measurement. This is a very significant contribution since it addresses the gaps of existing frameworks, as indicated by Potter & Beard (2010), Calder & Watkins (2010, 2012) Fomin et al. (2008), Susanto et al. (2012c, 2012h), that current existing frameworks do not provide a model for a formal readiness level measurement on how the ISMS standard is adopted by an organization.

ISF and ISM is an academic contribution to the scientific and practical environment. For future research, ISF could be made to accommodate and be customized to fit with other standards such as BS 7799, COBIT, ITIL, and others. ISF could possibly be implemented by other standards by following mapping stages through grouping of controls to the respective domains in each standard.

ISF is intended to introduce a novel algorithm for compliance measurement and investigation of ISMS as a bottom-up approach, designed to be implemented in high-level computer programming language, to produce a graphical user interface (GUI) that is easy to be used and powerful for ISO 27001 investigation. An innovative aspect of this approach is the development of a software (ISM) composed of two main functions: Security assessment management (SAM/e-assessment), which functions as log management and compliance reporting, and security monitoring management (SMM/e-monitoring) which functions as real-time monitoring for security-related events (further discussed in Chapter 6).

All those study contributions could be summarized as follows:

1. **A structured approach** for determining and mapping assessment issues, controls, clause and domain settings by **the new framework (ISF)** in order to organize security management issues in an ISMS standard (ISO 27001) effectively.

2. **A systematic mechanism for ISO 27001 refinement.** The refinement is used to verify and refine ISO 27001 to determine the degree of clarity of each essential control over its parameters. Refinement is a deterministic process, and since organizations have a number of information security controls, without refinement the controls tend to be somewhat disorganized and disjointed, having been implemented often as point solutions to specific situations or simply as a matter of convention. It is obvious that essential controls are very difficult to understand, immeasurable and difficult to be implemented by organizations and stakeholders (to be explained in detail in Chapter 4).

3. **ISM (integrated solution modeling software for RISC investigation).** The framework (ISF) has led us to develop ISM as a user interface between the stakeholder and ISF's approach to measuring information security awareness (ISA) and compliance level of the ISMS standard (ISO 27001) within an organization, such as protecting information and information systems from unauthorized access, use, disclosure, disruption, modification, perusal, inspection, recording or destruction. ISM consists of two major subsystems of e-assessment and e-monitoring. E-assessment is to measure ISO 27001 parameters based on the proposed framework with 21 essential controls and e-monitoring is to monitor suspected

activities that may lead to security breaches and provides real-time monitoring for security-related events. The software is equipped with a user record of accomplishment, functioning to determine users' patterns of assessment (Future explanation in Chapter 5).

## 1.6  BOOK STRUCTURE

This book is composed of seven chapters. Chapter 1 is the introduction which contains the background, problems and motivation of the research. This chapter also highlights the methodology employed and summarizes results and contributions of the book. Chapter 2 contains the literature review of the field of information security management systems, frameworks and managing risk associated with information security. Chapter 3 is concerned with the research methodology of the book. Chapter 4 discusses the proposed framework as a new approach to map security controls within six domains. Based on the framework we developed software (ISM) as a tool to measure readiness level with ISO 27001, discussed in Chapter 5. Chapter 6 illustrates testing and comprehensive evaluation conducted by ISM in respondent organizations and discussion of the result. Finally, Chapter 7 is the conclusion.

## 1.7  CONCLUDING REMARKS

The main aim of this study has been to map up the terrain of information security management in organizations. Securing information resources from unauthorized access is an extremely important, since information need to be managed in a proper and systematic manner as information security is quite complex. This research contributes a new approach for RISC investigations by offering a framework for the evaluation, formation and implementation of information security, through identifying ISMS basic building blocks (assessment issues, controls, clauses, and domains).

Practitioners and stakeholders can use the research's results (ISF, refinement, and ISM) presented here as blueprints for managing information security within their organizations. They can compare and benchmark their own processes and practices against these results and come up with new critical insights to aid them in their stages to information security standard (ISO 27001) adoption. Scholars in the field of information security management can use the existing results and build further on them to form a coherent and complete body of knowledge of the area.

Finally, an innovative aspect of this research is the proposed novel framework (ISF) and development of software (ISM). ISF enables the binding of organizational security policies and standards to the governance and compliance requirements. This contribution changes the landscape of information security standard adoption to a more structured approach and measurement. ISM is a semi-automated tool to assess the readiness of an organization to comply with ISO 27001 and subsequently assess the potential threats. ISM's two main functions of Security Assessment Management (SAM/e-assessment) and Security monitoring management (SMM/e-monitoring) could help an organization to review their circumstances regarding ISMS as the preliminary adoption stage.

## KEYWORDS

- integrated solution framework
- integrated solution modeling
- security assessment management
- software performance parameters

# LITERATURE REVIEW

## CONTENTS

### 2.1   INTRODUCTION

This chapter contains a series of references and guidelines for groundwork (Cooper, 1998) by grabbing and identifying possible gaps to consider the critical points of current knowledge including substantive findings that lead to a proposed novel theory, framework, formula, algorithm, and application-tool as an academic and practical contribution of the research to a particular topic (Dellinger, 2005; Hart, 2008). In agreement with Dellinger & Hart (2006) and Green & Hall (1984), its indicated that literature reviews unveil the current existing theories, findings and conclusions, and also identify the connectivity between existing works and the proposed study.

The motivation of this study is to reveal and determine the most challenging aspects of computer security (comsec), such as security breaches,

security awareness, and security hacks, as an organization's business support and enabler. Pipkin (2000) unveils "information security" from the
business standpoint and argues that information security should be an
integral part of a business. According to Sherwood et al. (2005), information security helps raise the trust of customers and partners towards the
organization that implemented it and allows the organization to effectively
use newly emerging technologies for greater commercial success (Cherdantseva & Hilton, 2013).

Information security enables business by increasing its competitiveness (Pipkin, 2000). Delving deeper into the business approach towards
information security, it should be understood that the security of information is required for businesses to improve efficiency through exploitation
of new technologies based on risk of business objectives and increase trust
from partners and customers. Other organizations also need security of
information for the same reasons (Dhillon, 2007; Easttom, 2012). According to Dhillon (2007) and Easttom (2011), the main aim of information
security is to protect business assets and support the achievement of business objectives.

Sherwood et al. (2005) addressed the change of information security
approaches in relation to the expansion of the hard perimeter of an organization caused by active collaboration, operation in a distributed environment and outsourcing of IT and other services. Consequently, information
security should not be considered purely from a technical perspective, but
also from a managerial, system architects' and designers' points of view
(Cherdantseva & Hilton, 2013).

There are many standards covering various aspects of information
security published by international organizations, national standards
bodies, non-profit organizations, and international communities (Cherdantseva & Hilton, 2013; Susanto et al., 2011a). The ISO 27001 series
of standards is an internationally recognized and widely adopted information security standard. The series was developed by a joint committee
between the International Organization for Standardization (ISO) and the
International Electronic Commission (IEC). ISO covers information security management, information security risk management, implementation
of information security management systems (ISMS), measurements and
metrics of ISMS. In 2000 ISO adopted BS7799, the standards published

by the British Standard Institute. This standard is based on the code of practice for information security management.

Business organizations need to seriously consider the security of their information systems, as secured information systems nowadays have a high impact on their successes. A secured information system not only supports their business activities but also enables companies to capture many opportunities created by e-business. Furthermore, secured information systems will increase the values of companies as information security is the main ingredient of trust in dealing with customers or clients online (Boyce & Jennings, 2002).

Many organizations face many challenges to make their information assets secure. To make sure those information assets are secured properly, proper guidelines provided by the information security standard need to be followed or complied with (ISACA, 2008, 2009; Von Solmn, 2005a, 2005b). This chapter aims to review important issues that will shape this research.

The remainder of this chapter describes and reviews the aspects of ISMS that relate to this research such as ISMS models, ISMS standards, ISMS frameworks, RISC[1] measurement approaches, software development methodologies, and SQ/SP[2] measurement.

## 2.2  COMPUTER SECURITY: TERM AND CONCEPT

Computer security (comsec) is a field and concept that covers all the processes and mechanisms by which computer-based equipment, information and services are protected from unintended or unauthorized users, access, change or destruction (Amoroso, 1994). Moreover, comsec can be grouped into logical and physical security. Logical security is defined as an action that protects data, information and programs in the system and physical security describes an action to prevent physical harm on the computer system hardware (Dlamini et al., 2009; Gollmann, 2010).

[1] Readiness and Information Security Capabilities – the RISC measurement is obtained from an investigation in stages to find out an organization's readiness and information security capabilities with respect to ISO 27001.
[2] Software Quality and Software Performance, which is the key performance and quality indicator of the ISM that is to undergo RISC testing. The data is generated in this phase to demonstrate the level of quality-performance, and user acceptance of the technology and features offered by the ISM.

Computer security is an area that offers an interesting aspect in the sense that there are conflicting objectives held by some of the actors of a single system, namely, attackers and legitimate users. It follows that depending on the goal that an actor is pursuing (attack or legal use), the use of a given computer system will differ dramatically. Whereas the roles of attackers are pretty clear (e.g., intrusion, denial of service), those of legitimate users regarding security are more subtle (Besnard & Arief, 2003).

Management commitment is a key success factor in implementing comsec. Management commitment is demonstrated through effective fostering of a computer security policy within the organization (Dlamini et al., 2009). There are many references and standards that provide guidelines on what to include or exclude in compiling a computer security policy (Fordyce, 1982).

Comsec decisions must have a set of understandings as their base, including: the management's risk acceptance, information as a vital resource, the forms of information and protection needs, the set of interchangeable security elements, and economics as the driving force behind personal and organizational processes (Schweitzer, 1982). The possible differences in views between the board of management and security managers become a cause of frequent tensions (Dlamini et al., 2009), since a security manager may blame an inability to obtain support for a really effective security program on management insensitivity to the real situation (Schweitzer, 1982). For instance, the lack of understanding is the result of management having a very broad view while the computer security specialist is narrowly focused and tends to see things in terms of fixed requirements.

## 2.2.1  ISSUES AND AREAS

PMC (2010) stated that based on technologies, trends, and issues, comsec is divided into four topics: network security, information security, infrastructure security, and cryptography-digital forensics.

*Network Security* consists of provisions and policies adopted by a network administrator to prevent and monitor unauthorized access, misuse, modification, or denial of a computer network. It involves the authorization of access to data in a network, which is controlled by the administrator.

*Information Security* is the practice of defending information from unauthorized access, use, disclosure, disruption, modification, perusal, inspection, recording or destruction. Governments, the military, corporations, financial institutions, hospitals, and private businesses amass a great deal of confidential information about their employees, customers, products, research and financial status. Most of this information is now collected, processed and stored in a digital format and transmitted across networks to other computers (Aceituno, 2005; Anderson, 2006; Dhillon, 2007;; Easttom, 2011; Lambo, 2006).

*Infrastructure Security* is the security measures taken to protect infrastructure, especially critical infrastructure, such as network communications, communication centers, server centers, database centers and IT centers. Infrastructure security seeks to limit vulnerability of these structures and systems from sabotage, terrorism, and contamination.

*Cryptography-Digital Forensics* is the process of encoding messages (or information) in such a way that eavesdroppers or hackers cannot read them and to examine digital media in a forensically sound manner with the aim of identifying, preserving, recovering, analyzing and presenting facts and opinions about the information (Halderman et al., 2009; Kruse & Heiser, 2002; Pieprzyk et al., 2003; Yasinsac et al., 2003).

## 2.2.2 PHENOMENON AND TRENDS

The development of ICT systems has become a critical component of globalization, shrinking both time and space, changing business environments, while on the other hand increasing the number of hackers and creating new vulnerabilities (Anderson, 2006; Dhillon, 2007). Criminals keep adapting their techniques to exploit vulnerabilities. As a result, cybercrime is becoming more common (Aceituno, 2005; Easttom, 2012), which implies that the number and cost of security breaches appear to be rising fast (Potter & Beard, 2010). Also, computer crimes cover a broad range of activities that include the work of computer hackers (some of whom have criminal intent) and those set to cause maximum disruption by producing and unleashing computer viruses.

Price Waterhouse Coopers (PWC) consultants estimated that the total cost of lost business and related security costs caused by the work of computer hackers and viruses was about £1 trillion. However, when viewed from an opportunity cost perspective it can also be argued, that in order for organizations to either make cost savings or provide higher returns to shareholders, managers are going to share information relating to the work of hackers and viruses produced by other organizations. Leaks of confidential information because of not being properly protected can result in inappropriate publicity for an organization. As a result, business may be lost or various investors may consider that further investment in the organization will be too risky and they may shift their money elsewhere (Anderson, 2001).

Potter and Beard (2012) in Information Security Breaches Survey, said *"The UK is under relentless cyber-attack and hacking is a rising risk to businesses. The number of security breaches large organizations are experiencing has rocketed and as a result, the cost to UK organizations of security breaches is running into the billions every year. Since most businesses now share data with their business partners across the supply chain, these numbers are startling and make uncomfortable reading for business leaders. Large organizations are more visible to attackers, which increases the likelihood of an attack on their own IT systems."*

Furthermore, Potter and Beard (2012) stated that 'incidents caused by staff' were experienced by 82% of large organizations in 2012. Unfortunately, no industry sector appears immune from these incidents. So far, telecommunications, utilities and technology companies appear to have the most reliable systems and the public sector and travel, leisure and entertainment companies are the most likely to have security problems (Figure 2.1).

Security incidents are getting more frequent. The average company can expect to have at least one security incident in a month while a large company might face it once a week. The total cost of the most serious security incident for small organizations in the UK ranged £10,000–£20,000 in 2008. This figure increased in 2010 to £27,000–£55,000. For large companies, the overall cost of the worst security incident was £90,000–£170,000 in 2008 and then increased to £280,000–£690,000 in 2010. In 2012, it was

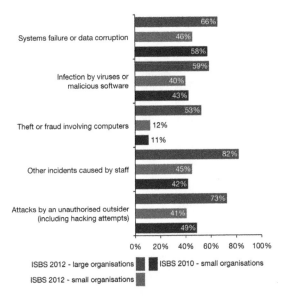

**FIGURE 2.1** What type of breaches did respondents suffer? (Potter & Beard, 2012 – Information Security Breaches Survey 2012).

about £15,000–£30,000 for small companies and £110,000–£250,000 for large companies (Table 2.1).

Pollitt (2005) indicated that 70% of IT budgets were spent on countering security breaches by providing new ways to protect customers from security threats through the concept of *"five principles of security"*: planning, proactive, protection, prevention and pitfalls. The five principles function to identify signs and flags of intruders in a network, establish guidelines for safeguarding user names and passwords and to match protection against physical access to information-technology facilities with the level of threat towards the security systems.

## 2.3 INFORMATION SECURITY: A PART OF COMPUTER SECURITY

Information security has become increasingly important in an era in which information is recognized as a key asset by many organizations. The rapid advancement of ICT and the growing dependence of organizations on ICT infrastructures continuously intensify interest in this discipline.

TABLE 2.1    What was the Overall Cost of an Organization's Worst Incident in the Last
Year? (Potter & Beard, 2012 – Information Security Breaches Survey 2012)

| | ISBS 2012 - small organisations | ISBS 2012 - large organisations |
|---|---|---|
| Business disruption | £7,000 - £14,000 over 1-2 days | £60,000 - £120,000 over 1-2 days |
| Time spent responding to incident | £600 - £1,500 2-5 man-days | £6,000 - £13,000 15-30 man-days |
| Direct cash spent responding to incident | £1,000 - £3,000 | £25,000 - £40,000 |
| Direct financial loss (e.g. loss of assets, fines, etc.) | £2,500 - £4,000 | £13,000 - £22,000 |
| Indirect financial loss (e.g. theft of intellectual property) | £4,000 - £7,000 | £5,000 - £10,000 |
| Damage to reputation | £100 - £1,000 | £5,000 - £40,000 |
| **Total cost of worst incident on average** | **£15,000 - £30,000** | **£110,000 - £250,000** |
| 2010 comparative | £27,500 - £55,000 | £280,000 - £690,000 |
| 2008 comparative | £10,000 - £20,000 | £90,000 - £170,000 |

Organizations pay increasing attention to information protection because
the impact of security breaches today have a more tangible, often devastat-
ing effect on business (Dlamini et al., 2009).

According to Shaw and Strader (2010), TJX[3] Companies lost from
36.2 to 94 million customers' credit and debit cards records in 2007. In
2011, Sony reported a data breach that resulted in the loss of personal
details of 77 million customers (Baker & Finkle 2011; Quinn & Arthur,
2011). According to the information security breaches survey 2010 (Potter
& Beard, 2010), the number of large companies in the UK that suffered
security incidents increased by 92% between 2008 and 2010. The total
cost of the worst security incident for large UK companies increased from
$170,000 in 2008 to $690,000 in 2010. In the US, between 2010 and 2011,
the number of security breaches detected by law enforcement increased by
33% (Trustwave.com, 2013). In 2012, security budgets increased worldwide compared to the previous year and predicted a stable 9% annual
growth of the security market until 2016. As a result, the worldwide

---

[3] The TJX Companies, Inc. is a US apparel and home goods company based in Massachusetts. It
claims to be the largest international retailer of apparel and home fashions. The company founded in
1956.

spending on security is expected to grow from $55 billion in 2011 to $86 billion in 2016.

In response to the growing interest, a significant amount of research has been conducted over the past two decades to cover various perspectives of information security that covered the technical side (Anderson, 2001a), the human factor (Lacey, 2011), the business and economic perspectives, and its governance (Anderson, 2001; Pipkin, 2000; Sherwood et al., 2005).

### 2.3.1 DEFINITION AND CONCEPT

This sub section contains a detailed analysis of the term "information security". Cherdantseva & Hilton (2013) explained information security in its different issues and views. First, an analysis of the term based on common English is conducted. Second, the definitions of the term as suggested in legal documents, and third, the understanding of information security among academics.

### 2.3.1.1 Information security: Common English

Cherdantseva & Hilton (2013) claim that formal or academic definitions are often distinct from the common comprehension of terms. Secure is defined as *"free from danger, damage, etc.; not likely to fail; able to be relied on"*. Precaution is defined as *"an action taken in advance to prevent an undesirable event"* (Parker, 1998).

The Oxford English Dictionary (2013) defines security as *"the state of being free from danger or threat"*. information security is a discipline, the main aim of which is to keep the knowledge, data and its meaning free from undesirable events, such as theft, espionage, damage, threat and other danger. information security includes all actions, taken in advance, to prevent undesirable events happening to the knowledge, data and its meaning so that the knowledge, data and its meaning could be relied on.

In general there are five points which should be highlighted in the definition of information security (Cherdantseva & Hilton, 2012): "first, there are no restrictions on the information type. In the broad sense, information security is concerned with information of any form or type (e.g., electronic, paper, verbal, visual). Second, information security includes all actions to protect information. Thus, information security is concerned not only with technical actions, but also deals with the full diversity of protecting actions required during information processing, storage or transmission. Third, the list of undesirable events is broad and open. The definition explicitly lists theft, espionage and damage of the information, but is not restricted to them. Thus, information security deals with the protection of information from all undesirable events. Fourth, the general definition of information security does not state any security goals such as confidentiality, integrity, availability or any other. Therefore, in line with the third point, the main aim of the discipline is the overall protection of information, and not just the achievement of several pre-defined security goals. Fifth, information security includes actions taken in advance. Therefore, information security should be concerned not only with an analysis of undesirable events, which have already taken place, but also with the anticipation of such events and an assessment of the likelihood."

### 2.3.1.2   Information Security: Legal Documents

Cherdantseva & Hilton (2013) stated that the ISO 27001 is a standard that is internationally recognized and widely adopted (Susanto et al., 2011a, 2011b). The series of rules covers information security management, information security risk management, measurements and metrics of information security management system (Alfantookh, 2009; Von Solms, 2005a, 2005b).

The National Information Assurance Glossary was published by the Committee on National Security Systems (CNSS) (CNSS, 2004). Although the standards were primarily oriented towards government systems, they are also useful for industry.

ISACA is a non-profit, global association of over 95,000 members world-wide and develops practices for information systems. ISACA is an originator of the globally accepted Control Objectives for Information and related Technology (COBIT) framework.

Cherdantseva & Hilton (2012) stated that CNSS and ISO define information security based on a set of security goals to be achieved. Thus, the essential discrepancy between the general comprehension of information security and the definitions provided in the standards is that by definition, information is secure if it is protected from all threats, whereas according to the standards, information is secure if it complies with the certain security goals (Von Solms, 2005a, 2005b).

The scope of information security defined by the ISO is wider than the scope defined by the CNSS. Apart from confidentiality, integrity and availability, the ISO also includes reliability, accountability, authenticity and non-repudiation in the realm of information security. For example, the breach of non-repudiation does not relate to any of the undesirable events stated in the CNSS definition. Although the set of security goals associated with information security in the CNSS and ISO standard vary, they agree that the three fundamental goals of information security are confidentiality, integrity and availability. ISACA clearly reflects this concept in its definition of information security (Table 2.2) (Cherdantseva & Hilton, 2012).

Since the standards correlate information security with a certain set of security goals, the provenance of the goals and their interpretations become extremely important. The series to define a list of security goals should be following these steps: (1) identify all possible threats to information; (2) categorize the threats; (3) define a security goal for each category of threats. However, due to the constant change in the environment, new threats constantly emerge and information received at the first step quickly becomes obsolete (Cherdantseva & Hilton, 2012; Susanto & Almunawar, 2011a).

**TABLE 2.2** Definitions of Information Security and Integrity Legal Documents (Cherdantseva & Jeremy Hilton, 2013)

| Legal Document | Definition |
| --- | --- |
| ISO | Preservation of confidentiality, integrity and availability of information. Note, in addition, other properties, such as authenticity, accountability, non-repudiation and reliability can also be involved. |
| CNSS | The protection of information and information systems from unauthorized access, use, disclosure, disruption, modification, or destruction in order to provide confidentiality, integrity, and availability. |
| ISACA | Ensures that only authorized users (confidentiality) have access to accurate and complete information (integrity) when required (availability). |

Parker (1998) indicated the definitions provided in the standards are used to define an organization's information security program, strategy and policies. Any shortcomings of the information security standards in this context will lead to undesirable consequences that stem from over-looking essential threats and critical vulnerabilities that stay below the radar of information security.

In comparison, the definitions of the documents discussed as shown previously in Table 2.2 narrow down the scope of the discipline because they define confidentiality, integrity and availability as the fundamental goals of information security, rather than an overall protection of informa-tion. It is obvious now that the ways a system can fail could necessarily be characterized by a breach of confidentiality, integrity or availability.

### 2.3.1.3   Information Security: An Academic Perspective

Cherdantseva & Hilton (2012) described that since the late 1970s, informa-tion security has been rigorously associated with the triad factors of con-fidentiality, integrity and availability, so-called the CIA-triad (Whitman & Mattord, 2011). There is a pronounced tendency to extend the scope of information security beyond the CIA-triad since the latter is found to be no longer adequate (Parker, 1998; Whitman and Mattord, 2011) for a complex interconnected environment. A plethora of security goals are considered relevant to information security and intensively discussed in the literature. Table 3 provided below lists the security goals associated with the discipline as illustrated in the security-related publications (Cher-dantseva & Hilton, 2012).

Parker (1998) criticized how the information security definitions were being limited to the CIA-triad and claims them being dangerously incor-rect. Parker introduces a new model of information security that consists of six foundation elements: confidentiality, integrity, availability, possession or control, authenticity and utility. He argues that his model replaces the incomplete description of information security limited to the CIA-triad. Albeit the model of information security suggested by Parker (1998), is not widely accepted, the research undertaken is fruitful because it addresses three issues, essential for the clarification of information security:

1. The focus of the discipline is set on protection of information, rather than on protection of an information system. Parker consistently includes in his model properties of information and does not mix them with security mechanisms.
2. The importance of a complete and accurate definition of the discipline and, consequently, of the discipline's goals is highlighted and justified.
3. An attempt to extend the model of information security and to address the limitations of the CIA-triad is undertaken. Looking beyond the CIA-triad leads to a broadening of information security from a technical to a multidimensional discipline.

Anderson (2001) confirmed that information security is more than the CIA-triad. Anderson proclaims a multidimensional approach to information security and sets forth that people, institutional and economic factors are no less important than the technical ones. Anderson (2001, 2006) conducted an analysis of economic incentives behind some information security failures and concluded that a purely technical approach to information security is ineffective. Further, Anderson stated that collaboration between managers, economists and ICT staff is required in order to solve problems related to information security. While Anderson (2006) provided the general inside view on the economic incentives behind information security, Gordon and Loeb (2002) looked at the economics of investments into information security.

In 2002, Gordon proposed an economic model that helped to determine the optimal amount of investment in information securitty. Gordon and Loeb associated information security with such goals as the confidentiality, availability, authenticity, non-repudiation and integrity of information (Gordon and Loeb, 2002). The importance of economic motives was also recounted by Schneier (2008), who stated that the number of vulnerabilities may only be reduced "*when the entities that have the capability to reduce those vulnerabilities have the economic incentive to do so*". In addition to economics, Schneier revealed that consideration of psychology and management to be essential for information security (Schneier, 2008).

Schneier (1999, 2001) described information security as a process that includes understanding of threats, design of polices and building of countermeasures to address the threats and further stated that all the components of the process must fit together in order to create the best state of the overall process. He distinguished the following goals of information security: privacy, information classification that was referred to as multilevel security, anonymity, authentication, integrity and audit (Schneier, 2009). Schneier listed among the security goals not only properties of information, but also security mechanisms or abilities of information systems (e.g., authentication).

Cherdantseva & Hilton (2012) and Pipkin (2000) defined information security as a process, in this case as *"the process of protecting the intellectual property of an organization"*. Pipkin included and discussed in detail ten security goals in the scope of information security: awareness, access, identification, authentication, authorization, availability, accuracy, confidentiality, accountability and administration. This is another confirmation of a wide trend in information security to combine security goals and security mechanisms as a result of considering information and information systems simultaneously to be subjects of protection in information security (Cherdantseva & Hilton, 2012) (Table 2.3).

## 2.3.2  INFORMATION SECURITY AWARENESS

Information security awareness (ISA) is the knowledge and attitude of members of an organization regarding the protection of tangible and intangible assets, especially information assets (Furnell, 2005). According to the European Network and Information Security Agency (ENISA, 2012), ISA is an awareness of the risks and available safeguards in the first line of defense for the security of information systems. The focus of security awareness should be to achieve a long-term shift in the attitude of employees, promoting a cultural and behavioral change within an organization towards information security.

Previous studies on ISA have highlighted a number of important topics such as information security effectiveness (Kankanhalli et al., 2003; Straub, 1990; Woon and Kankanhalli, 2007), security planning and risk management (Hoo, 2000; Straub and Welke, 1998), the economics of information security and evaluation of information security investments (Cavusoglu et al., 2004a, 2004b, 2004c), and the design, development, and alignment

**TABLE 2.3** Analysis of the Literature in Terms of Goals Associated with Information Security (Cherdantseva & Jeremy Hilton, 2013)

| Reference(s) | Confidentiality | Integrity | Availability | Accountability | Assurance | Authentication | Non-Repudiation | Authenticity | Reliability | Effectiveness | Efficiency | Compliance | Utility | Control | Authorization | Awareness | Access | Identification | Accuracy | Administration | Information Classification | Anonymity | Audit | Safety |
|---|---|---|---|---|---|---|---|---|---|---|---|---|---|---|---|---|---|---|---|---|---|---|---|---|
| Clark & Wilson, 1987 | X | X | X | | | | | | | | | | | | | | | | | | | | | |
| McCumber, 1991 | X | X | X | | | | | | | | | | | | | | | | | | | | | |
| Parker, 1998 | X | X | X | | | | | X | | | | | | X | | | | | | | | | | |
| Pipkin, 2000 | X | | X | X | | | | | | | | | | X | | | | | | | | | | |
| Schneier, 2001 | | X | | | | X | | | | | | | | | | | | | | | | | | |
| Gordon & Loeb, 2002 | X | X | X | | | | X | X | | | | | | | | | | | | | | X | X | |
| Avizienis et al., 2004 | X | X | X | X | | | | | X | | | | | | | | | | | | | X | X | |
| ISO, 2004 | X | X | X | | | | X | X | X | | | | | | X | X | X | X | X | X | X | | | |
| ITGI, 2007 | X | X | X | | | | | | X | X | X | X | | | X | X | X | X | X | | X | | | |
| ISACA, 2008 | X | X | X | | | | | | | X | X | X | | | X | X | X | X | X | | | | | |
| CNSS, 2004 | X | X | X | | | X | X | | | | | | | | X | X | X | X | | | | | | |
| Tiller, 2010 | X | X | X | | | | | | | | | | | | X | X | X | | | | | | | |

**TABLE 2.3** (Continued)

| Reference(s) | Confidentiality | Integrity | Availability | Accountability | Assurance | Authentication | Non-Repudiation | Authenticity | Reliability | Effectiveness | Efficiency | Compliance | Utility | Control | Authorization | Awareness | Access | Identification | Accuracy | Administration | Information Classification | Anonymity | Audit | Safety |
|---|---|---|---|---|---|---|---|---|---|---|---|---|---|---|---|---|---|---|---|---|---|---|---|---|
| Dubois et al., 2010 | X | X | X | X | | | X | | | | | | | | | | | | | | | | | |
| Whitman & Mattord, 2011 | X | X | X | | | | | X | | | | | X | X | | | | | X | | | | | |
| Alfantookh, 2009 | X | X | X | | | | | | | | X | X | | | | | | | | | | | | |
| Al-Osaimi et al., 2008 | X | X | X | | | | | | | | X | X | | | | | | | | | | | | |
| Von Solms, 2005a, 2005b | X | X | X | | | | | | | X | X | | | X | | | | | | | | | | |

of the ISP (Doherty and Fulford, 2006; Siponen, 2006). While these studies have expanded the understanding of information security from various perspectives, ISA research is particularly underrepresented in the leading information security journals (Siponen and Willison, 2007). An emerging research stream on the human perspective of information security focuses on end-user (insider) behaviors and attempts to identify the factors that lead to compliance behavior regarding information security. The current literature recognizes that *insiders*, a term that refers to employees who are authorized to use a particular system or facility (Neumann, 1999), may pose a challenge to an organization because any ignorance, mistakes, and deliberate acts can jeopardize information security (Bulgurcu et al., 2009, 2010; Lee et al., 2003).

ISA should be viewed as a key enabler and an integral part of a business, not as a series of rules restricting the efficient working of business processes. Pipkin (2000) and Sherwood et al. (2005) unveiled ISA from the business standpoint and argued a need for information security to become a business enabler and an integral part of a business, and that ISA may help to raise trust of an organization by customers and partners, and to allow an organization to effectively use newly emerging technologies for greater commercial success. Therefore, it would be prudent to support the information assets by trying to stop the information breaches and overcome several issues associated with ISA. Peltier (2005a, 2005b) and Furnell (2005) indicated several items related to ISA as follows:

- The important and sensitive information and physical assets, such as trade secrets, privacy concerns and confidential classified information.
- Employees' and contractors' responsibilities in handling sensitive information.
- Requirements for proper handling of sensitive material in physical form.
- Proper methods for protecting sensitive information on ICT systems; password policy and use of factor authentication.
- Other computer security concerns: malware, phishing, social engineering.

- Workplace security: building access, wearing of security badges, reporting of incidents, forbidden articles, etc.
- Consequences of failure to properly protect information: potential loss of employment, economic consequences to the firm, damage to individuals whose private records are divulged, and possible civil and criminal penalties.

Boss et al., (2009) introduced the concept of *mandatoriness*, which has been shown to motivate individuals to take security precautions. Despite the importance of ISA, there is a paucity of empirical studies that analyze the impact of ISA on information security. Siponen (2006) conceptually analyzed ISA and suggested methods to enhance awareness. A few conceptual studies (Furnell et al., 2006; Hentea, 2005; Thomson and Von Solms, 1998) have highlighted the importance of ISA education and training. Puhakainen and Ahonen (2006) proposed a design theory for improving ISA campaigns and training. D'Arcy et al. (2009) suggested that organizations can use three security countermeasures—user awareness of security policies; security education, training, and awareness programs; and computer monitoring—to reduce user's misuse. Beyond showing the direct influence of ISA on an employee's attitude towards compliance, the countermeasures aim to understand the antecedents of compliance by disentangling the relationships between ISA and an employee's outcome beliefs about compliance and noncompliance. For instance, ISA issues within organizations were apparent in the report by Potter and Beard (2012), where it stated that '*incidents caused by staff*' were experienced by 82% of the sampled large organizations.

However, ISA implementation should be viewed as one of an organization's serious efforts to improve corporate selling point to customers (Kotler, 2002), as corporate imaging and branding (Dwyer, 1987), to win the competitive edge within its related business area (Morrison, 2003), as one of the marketing tools (Kottler, 2002), to increase corporate profitability as an effect of customers' trust and also to create loyal customers through trust (Anderson, 2001; Brown, 2000). Kottler (2002) stated, "*it is obvious that the business organization as producer, are interdependent with their loyal customers for the business sustainability*". Customer loyalty is all about attracting the right customers, winning their trust and providing convenience, getting them to buy, buy often, buy in higher quantities, and bring even more customers (Almunawar et al., 2013; Dick & Basu, 1994).

## 2.3.3  RISK MANAGEMENT ASSOCIATED WITH INFORMATION SECURITY

Furnell and Karweni (1999) examined the general requirement for information security technologies in order to provide a basis for trust in the ICT domain. In a modern business environment, ICT is strongly recognized and in fact modern businesses are hardly operated without ICT (Solis, 2012). As a result, modern business organizations are highly dependent on ICT either as a support in their operations or as a business enabler (Pipkin, 2000; Sherwood et al., 2005). The dependency of business organizations on ICT makes the issue of information security very important to address (Baraghani 2008; Susanto et al., 2012b) as this issue has become a main concern for customers (Ramayah et al. 2003). Therefore, an organization needs to assure information resources, electronic services and transactions are well protected at the acceptable level of the information security risk through risk management (dealing with information security) or information risk management (IRM) (D'Arcy and Brogan, 2001; Wu and Olson, 2009).

Information risk management (IRM) is one component of ISO 27001, it is mentioned in the clause of "*business continuity management*", and falls under the item "*information security in the business continuity process and business continuity risk management/assessment*" (ISO, 2005). IRM is a recurrent activity that deals with the analysis, planning, implementation, control, and monitoring of measurements and the enforced security policy, procedures and practices to reduce the effects of risk to an acceptable level according to ISO 27001 (Blakley, 2001; Lichtenstein & Williamson 2006).

IRM consists of methods and processes used by organizations to manage risks and to seize opportunities related to the achievement of their objectives. By identifying and proactively addressing risks and opportunities, an organization protects and creates value for their stakeholders, including owners, employees, customers, regulators, and the society overall. IRM has gained considerable importance in organizations because of the increased regulatory demands and a growing awareness of its importance in preventing systemic failure. In other words, companies must demonstrate that they actually use IRM as a key component of business processes (Wu & Olson, 2009).

## 2.3.4   BUSINESS BREAKTHROUGH

As mentioned in the previous section, Pipkin and Sherwood discussed information security from the business standpoint and argued a need for information security to become a business enabler and an integral part of a business that may help to raise trust of an organization by customers and partners, and to allow an organization to effectively use newly emerging technologies for a greater commercial success.

Sherwood et al. (2005) adopted a multidimensional and enterprise-wide approach to information security and included in the scope of information security, for example, marketing and customer service. Cherdantseva & Hilton (2013) declared protection of business assets and assistance of the achievement of business goals to be the main aim of information security. Pipkin (2000) and Sherwood et al. (2005), by the adoption of a business-oriented approach, supported the tendency to extend the realm of the discipline. Thus, information security is no longer considered purely from a technical perspective, but also from a managerial, system architect's and designer's points of view. Von Solms (2001), Pipkin (2000), and Sherwood et al. (2005) confirmed the transition of information security from the purely technical to the multidimensional discipline and identified 13 closely interdependent dimensions of information security:

1.  The Strategic/Corporate Governance Dimension;
2.  The Governance/Organizational Dimension;
3.  The Policy Dimension;
4.  The Best Practice Dimension;
5.  The Ethical Dimension;
6.  The Certification Dimension;
7.  The Legal Dimension;
8.  The Insurance Dimension;
9.  The Personnel/Human Dimension;
10. The Awareness Dimension;
11. The Technical Dimension;
12. The Measurement/Metrics (Compliance monitoring/Real time IT audit) Dimension;
13. The Audit Dimension.

The list of the information security dimensions may be extended with the following dimensions derived from the comparative analysis of Anttila et al. (2004) and Shoemaker et al. (2004):

1. The Physical Security Dimension;
2. The System Development Dimension which ensures that the security is built into the development process;
3. The Security Architecture Dimension;
4. The Business Continuity Dimension;
5. The Privacy Dimension.

The shift of information security from the technical to the broad, multidimensional discipline is also supported by Lacey (2009), who recounted that information security *"draws on a range of different disciplines: computer science, communications, criminology, law, marketing, mathematics and more"*. Lacey (2010, 2011) indicated the importance of technologies for protection of information, and emphasizes the importance of the human factor which is based on the fact that all technologies are designed, implemented and operated by people. In addition to the human factor, Lacey also considered how organizational culture and policies affect information security.

At a time of interconnection and collaborative information sharing, information security becomes more challenging since information now needs protection not only within the safety of the organization's perimeter, but also between organizations (Cherdantseva et al., 2011; Lacey, 2010, 2011; Pipkin, 2000; Sherwood et al., 2005). The multidimensional nature and the broadening scope of information security also identified that three areas have become important for information security: legal and regulatory compliance (standards), risk management and information security management systems (ISMS).

As a consequence, the number of people involved in information security is increasing. If previously there were only technical experts involved in information security, at present managers, legal personnel, compliance regulators and human resources specialists are also involved in information security. Tiller (2010) stated that information security, most importantly, in addition to a comprehensive approach is required to be agile and adaptable to meet the requirements of continuously evolving business needs (Pipkin, 2000; Tiller, 2010; Von Solms, 2001).

Nowadays, it is clear that the technology alone is insufficient for solving complex tasks of the discipline. Business needs, the human factor, economic incentives, cultural and organizational aspects should be taken into account in order to achieve an adequate protection of information. At present a comprehensive, multidimensional approach to the protection of information is required. Below is a list of recent trends in the discipline of information security (Cherdantseva et al., 2009; Cherdantseva & Hilton, 2013):

1.  The information security moves from a low-level technical activity and responsibility of computer specialists to a top priority activity dealt with at the managerial level (Dlamini et al., 2009).
2.  The information security becomes a multidimensional discipline, and several issues related to management (Pipkin, 2000; Sherwood et al., 2005; Tiller, 2010), marketing (Sherwood et al., 2005), economics (Anderson, 2001; Schneier, 2008), physiology (Lacey, 2011; Schneier, 2008), law (Lacey, 2010, 2011; Von Solm, 2001), sociology (Theoharidou et al., 2005), criminology (Lacey, 2011; Theoharidou et al., 2005), mathematics (Anderson, 2001, 2003, 2006; Lacey, 2010) and other disciplines are now in the scope of information security.
3.  Information security shifts from the protection of closed IT systems to the protection of open connectivity systems used for collaboration and interconnection, within and between organizations (ISACA, 2008, 2009; Pipkin, 2000; Sherwood et al., 2005).
4.  The CIA-triad is considered to be an important scope of information security, but not reflecting the complete issues of information security (Anderson, 2001; Parker, 1998).

### 2.3.5   CHALLENGES

A major threat and challenges to organizational information security is the rising number of incidents caused by social engineering attacks. Social engineering is defined as the use of social disguises, cultural ploys, and psychological tricks toward computer users for the purpose of information gathering, fraud, or gaining computer system access (Anderson & Adey, 2011). Despite the continued efforts of organizations to improve user awareness about information security, social engineering malware has been successfully spread across the Internet and have infected many computers (Bailey et al., 2007; Bakos, 1991; Blyth & Kovacich, 2001).

Botnet (robot network)[4] leverages a wide range of malware to infect network-accessible devices, with the majority of the devices being personal computers in homes, businesses, schools, and governments. Once infected, these devices (or nodes) form botnets and are remotely controlled by the botmasters for illicit activities such as sending e-mail spam and extortion by threats of launching distributed denial-of-service (DDoS) attacks (Dietrich et al., 2013; Khosroshahy, 2013). To succeed, a social engineering activity such as malware needs to be activated and run on the system. Identifying attack strategies is vital to develop countermeasures that can be incorporated into preventive mechanisms like e-mail filtering and end-user security. Information on the behavior of the malware during propagation helps in the creation of early warning systems (Rahim & Muhaya, 2010). When computer malware is activated, it makes various changes in the computer by opening backdoors that enable it to spread to other machines. It also executes defensive strategies in order to remain undetected. Identification for malware activation is helpful in discovering malware activity in its early stages on end-user machines, and blocking it from being executed completely and propagating further (Rahim & Muhaya, 2010).

There has been an explosion of malware appearance on computing systems over the last decade, whose goal is to compromise the confidentiality, integrity and availability (CIA-triad) of infected computing systems. The exponential growth in malware population is articulated by pertinent encyclopedias: for example, by March 2009, there were 340,246 specimens of malware listed in TrendMicro (2011); of which, only 37,950 species were collected before 2000 and 123,802 breeds were compiled in 2006 (Trend Micro, 2011).

Malware often locates victim systems using an array of mechanisms including host scanning, hit-list gleaning, and network snooping. In the process, affected systems are bombarded by voluminous network traffic and a large amount of their computational resources are consumed, thus

[4] Botnet is a collection of Internet-connected programs communicating with other similar programs in order to perform tasks. This can be as mundane as keeping control of an Internet Relay Chat (IRC) channel, or it could be used to send spam email or participate in distributed denial-of-service attacks. The word botnet is a stand for robot and network.

decreasing their productivity. When successful, the malware frequently implants in infected machines extra pieces of code, termed "payloads", that typically lead to information leakage, distributed denial-of-service attacks, and security degradation (Zeltser et al., 2003). To extend its life span, malware not only cover its code with cryptographic techniques, but also resort to polymorphism and metamorphism (Kawakoya et al., 2010; Leder et al., 2009). In this regard, each of the produced clones diversifies in terms of code sequence and functionality, defeating anti-malware products that exclusively rely on pattern matching for malware detection (Costa & D'Amico, 2011; Walenstein et al., 2010).

Malicious software is an extreme threat to the network ecozystem as it can irreparably damage computing systems. The essential challenge in identifying malware species is to capture the characteristics that distinguish them. Malware traces left on infected machines include modifications to the file system, manipulations on registry databases, or network activities, which are typically obtained by static analysis of malicious binaries (Shieh and Gligor, 1997). The analysis results of malware species are usually assembled into an encyclopedia – a comprehensive account of malware (Shieh and Gligor, 1997). Trend Micro's encyclopedia listed more than 340,246 by the end of last decade (Trend Micro, 2011).

## 2.4  INFORMATION SECURITY MANAGEMENT SYSTEM STANDARD

A regulation that governs information security within an organization is called an information security management system (ISMS). An ISMS is a set of policies concerned with information security management or IT related risks, consisting of policies concerned with information security management (Alfantookh, 2009; Saleh et al., 2007a, 2007b; Von Solms, 2001). Pipkin (2000) and Sherwood et al. (2005) claimed that information security has become a business enabler and an integral part of a business, therefore as a consequence organizations pay increasing attention to information protection by implementing an ISMS standard.

The principle behind an ISMS standard is that organizations should design and be aware of information security scenarios; implement and maintain a coherent set of policies and processes to manage risks, vulnerabilities, and threats to its information assets. An ISMS standard must remain effective and efficient in the long-term, adapting to changes in the internal organization and external environment. An ISMS standard normally incorporates the typical "Plan-Do-Check-Act" (PDCA) or Deming cycle (Madu & Kuei, 1993) (Figure 2.2) approach:

- *The Plan phase* is about designing the ISMS, assessing information security risks and selecting appropriate controls.
- *The Do phase* involves implementing and operating the controls.
- *The Check phase* is to review and evaluate the performance (efficiency and effectiveness) of the ISMS.
- In *the Act phase*, changes are made where necessary to bring the ISMS back to peak performance.

There are several guidelines within ICT Governance which lead to information security such as PRINCE2, OPM3, CMMI, P-CMM, PMMM, ISO 27001, BS7799, PCIDSS, COSO, SOA, ITIL, and COBIT (Eloff & Eloff, 2005; Susanto et al., 2011a, 2011b; Von Solms, 2005). Unfortunately, some of these standards are not well adopted by the organizations, for a variety of reasons. We filter those standards into five main ISMS standards: ISO 27001, BS 7799, PCIDSS, ITIL, and COBIT. The review of each standard and a comparative study was conducted to determine their respective strengths, focuses, main components and their adoption

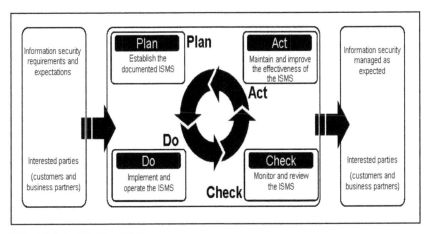

**FIGURE 2.2**   PDCA – ISMS.

and implementation (Susanto et al., 2011a, 2012a). Further discussion in Section 2.4.1 summarizes the facts.


### 2.4.1 ISO 27001

ISO was founded on February 23, 1947, to promulgate worldwide propri-etary industrial and commercial standards. Its headquarter is in Geneva, Switzerland and it has 163 national members out of the 203 total countries in the world (Kosutic, 2013). The international standard of ISO 27001 specifies the requirements for establishing, implementing, operating, monitoring, reviewing, maintaining and improving an ISMS within an organization (Alfantookh, 2009; Saleh et al., 2007a). It was designed to ensure the selection of adequate and proportionate security controls to pro-tect information assets. ISO 27001 is a member of ISO 27K standards. ISO 27K standards are being actively developed, the content, scope and titles of standards often change during the drafting and approvals process. ISO 27K standards are consists of:

1. ISO/IEC 27000:2014. **Information Security Management Sys-tems – Overview and Vocabulary.** The vocabulary or glossary of carefully worded *formal definitions* covers most of the specialist information security-related terms in the ISO27K standards. For an example, information security (such as "risk") has different meanings or interpretations according to the context of the author's intention and the reader's perception.

2. ISO/IEC 27001:2013. **Information Security Management Sys-tems – Requirements.** ISO/IEC 27001 formally specifies the requirement of Information Security Management System (ISMS), a suite of activities concerning the management of information security risks. The ISMS is an overarching management framework through which the organization identifies, analyzes and addresses its information security risks. The ISMS ensures that the security arrangements are fine-tuned to keep pace with changes to the secu-rity threats, vulnerabilities and business impacts.

3. ISO/IEC 27002:2013. **Code of Practice for Information Secu-rity Management.** ISO/IEC 27002:2013 is a code of practice - a

generic, advisory document, not a formal specification such as ISO/ IEC 27001. It recommends information security controls addressing information security control objectives arising from risks to the confidentiality, integrity and availability of information.

4. **ISO/IEC 27004:2009. Information Security Management – Measurement.** ISO/IEC 27004:2009 provides guidance on the development and use of measures and measurement in order to assess the effectiveness of an implemented information security management system (ISMS) and controls or groups of controls, as specified in ISO/IEC 27001. This would include policy, information security risk management, control objectives, controls, processes and procedures, and support the process of its revision, helping to determine whether any of the ISMS processes or controls need to be changed or improved.

5. **ISO/IEC 27007. Guidelines for Information Security Management Systems Auditing (focused on the management system).** ISO/IEC 27007 provides guidance for accredited certification bodies, internal auditors, external/third party auditors and others auditing ISMSs against ISO/IEC 27001. ISO/IEC 27007 reflects and largely refers to ISO 19011, the ISO standard for auditing quality and environmental management systems – "management systems" It is also draws on ISO 17021 *Conformity Assessment – Requirements for bodies providing audit and certification of management systems* and aligns with ISO/IEC 27006, the ISMS certification body accreditation standard.

6. **ISO/IEC TR 27008. Guidance for Auditors on ISMS Controls (focused on the information security controls).** This standard provides guidance for all auditors regarding information security management systems controls selected through a risk-based approach for information security management. It supports the information security risk management process and internal, external and third-party audits of an ISMS by explaining the relationship between the ISMS and its supporting controls. It provides guidance on how to verify the extent to which required "ISMS controls" are implemented. Furthermore, it supports any organization using ISO/IEC 27001 and ISO/IEC 27002 to satisfy assurance

requirements, and as a strategic platform for information security governance.

7. ISO/IEC 27014. **Information Security Governance.** The standard provides guidance on concepts and principles for the governance of information security, by which organizations can evaluate, direct, monitor and communicate the information security related activities within the organization" and is "applicable to all types and sizes of organizations.

These standard is usually applicable to all types of organizations (e.g., commercial enterprises, government agencies, and non-profit organizations) and all sizes from micro-businesses to huge multinationals. Moreover, ISO 27001 is also intended to be suitable for several different types of functionality and usability, including identification and clarification of existing information security management processes:

1. The usage by the management of organizations to determine the status of information security management activities;

2. The usage by the internal and external auditors of organizations to demonstrate the information security policies, directives and standards adopted by an organization and determine the degree of compliance with those policies, directives and standards;

3. As functions as provided relevant information about information security policies, directives, standards and procedures to trading partners and other organizations that they interact with for operational or commercial reasons (Furnell et al., 2006);

4. Implementation of a business enabling information security; and use by organizations to provide relevant information about information security to customers (Calder & Watkins, 2010).

### 2.4.1.1   Key Benefits of ISO 27001 Implementation

By adopting ISO 27001, organizations have the opportunity to prove credibility and show customers and other stakeholders that the organization is working according to recognized best practices. This credibility is often a deciding factor, giving the certified organization a competitive advantage (an extremely valuable intangible asset) (RMST, 2012). In today's competitive market, more organizations are adopting ISO 27001, resulting in a paradigm shift in the requirements for organizations whose businesses are related to information security (Kosutic, 2013). Customers are beginning to make ISO 27001 a requirement for suppliers to comply with, thus guaranteeing that suppliers are following the best practices. Kosutic (2012) stated several benefits of ISMS, especially the adoption of ISO 27001. The following are four of the most important of those benefits:

*Compliance.* It often shows the quickest "return on investment" – if an organization complies with various regulations regarding data protection, privacy and IT, then ISO 27001 can bring in the methodology which enables to do it in the most efficient way.

*Marketing edge.* In a competitive market, it is sometimes very difficult to find something that will differentiate as the unique selling point of an organization in the viewpoint of customers. ISO 27001 could be indeed a unique selling point, especially if an organization handles clients' sensitive information.

*Lowering the expenses.* Information security is usually considered as a cost with no obvious financial gain. However, there is financial gain if an organization can minimize expenses caused by incidents. There is no methodology to calculate how much money could be saved if an incident is prevented. It always sounds good if an organization can bring such cases to customers' attention.

*Putting a business in order.* An organization which has been growing sharply might be facing problems such as who has to decide what, who is responsible for certain information assets, who has to authorize access to information systems, etc. ISO 27001 is particularly good in sorting these things out – it will force an organization to define very precisely both the responsibilities and duties, and therefore strengthen an organization's internal information security.

### 2.4.1.2 Assessment Stages

An ISO 27001 assessment is probably the most complex part of ISO 27001 implementation (Kosutic, 2010, 2013) but at the same time it is the most important step of information security adoption. It sets the foundation for information security in an organization, as the main philosophy of ISO

27001 is to find out which incidents could occur (i.e., assess the risks) (Alfantookh, 2009) and then find the most appropriate ways to avoid such incidents by referring to the controls associated with them (i.e., treat the risks) (Calder & Watkins, 2012). The following are six basic steps in ISO 27001 assessments (Alfantookh, 2009; Calder & Watkins, 2012; Kosutic, 2010, 2013):

*Risk Assessment Methodology.* An organization needs to define rules on how to perform risk management since the biggest problem with assessment happens if different parts of the organization perform it in different ways. Therefore, an organization needs to define whether to perform qualitative or quantitative risk assessment. The scale should be well-defined if organizations choose the qualitative risk assessment. It is also important to define the acceptable level of risk.

*Risk Assessment Implementation.* Once the rule is fixed, an organization can start finding out which potential problems could happen, list all assets, then the threats and vulnerabilities related to those assets. They must assess the impact and likelihood for each combination of assets/threats/vulnerabilities and finally calculate the level of risk.

*Treatment Implementation.* This stage is where an organization needs to find a solution to implement, for instance, with minimum investment.

*ISMS Assessment Report.* An organization needs to document everything that has happened or was performed, for the purpose of audit and review.

*Statement of Applicability (SoA).* This stage shows the security profile of an organization. Based on the results of the risk treatment and risk management, an organization needs to list all the controls that have been implemented. This document (SoA) is very important because the certification auditor will use it as the main guideline for the audit. This SoA in detail will be explained in Subsection 2.3.1.4.

*Risk Treatment Plan.* To define exactly who is going to implement each control, including the timeframe and budget, the organization prepares a document called an 'Implementation Plan' or 'Action Plan'. The details of controls are discussed in the following subsection.

## 2.4.1.3   Assessment Issues, Controls, and Clauses

Information security is characterized as a set of policies consisting of 11 clauses or objectives, some of them containing essential controls (ECs). ECs are the most important controls and are concerned with the first security level within standards (Figure 2.3), (Alfantookh, 2009; Bakry, 2009). All those clauses are listed as follows (ISO, 2005; Saleh et al., 2007a, 2007b):

1.  Security Policy *(1 essential control);*
2.  Organizing information security *(1 essential control);*
3.  Asset Management security*;*
4.  Human Resources security *(3 essential controls);*
5.  Physical and Environmental security*;*
6.  Communication and Operations Management*;*
7.  Access Control*;*
8.  InformZation Systems Acquisition, Development and Maintenance *(5 essential controls);*
9.  Information Security Incident Management *(3 essential controls);*
10. Business continuity Management *(5 essential controls);*
11. Compliance *(3 essential controls);*

Moreover, ISO 27001 standard contains 21 ECs, as follows (Alfantookh, 2009; Kosutic, 2010, 2013; Saleh et al., 2007a, 2007b; Susanto et al., 2011c, 2012a) (Table 2.4): (1) document, (2) input data validation, (3) control of internal processing, (4) message integrity, (5) output data validation, (6) control of technical vulnerability, (7) allocation of information security responsibilities, (8) responsibilities and procedures, (9) learning from information security incidents, (10) collection of evidence, (11) information security in the business continuity process, (12) business continuity and risk assessment, (13) developing and implementing continuity plans including information security, (14) business continuity planning framework, (15) testing, maintaining and re-assessing business continuity plans, (16) management responsibilities, (17) information security awareness, education and training, (18) disciplinary process, (19) intellectual property rights, (20) protection of organizational records, and (21) data production and privacy of personal information.

### 2.4.1.4    Mandatory Document and Statement of Applicability

The Statement of Applicability (SoA) is the central document that defines how an organization will implement (or has implemented) information security. SoA is the main link between the risk assessment and treatment and the implementation of information security to define which of the suggested 133 controls (including the 21 essential controls of security measures) will apply, and how to implement it. Actually, if an organization goes for the ISO 27001 certification, the certification auditor will take the SoA and check around the organization on whether it has implemented the controls in the way described in the SoA (Alfantookh, 2009; Kosutic, 2010, 2013;; Saleh et al., 2007a, 2007b).

Moreover, the adoption of the standard requires four documented procedures (Kosutic, 2013; Saleh et al., 2007b): *first,* the *procedure for the control of documents* (document management procedure) should define who is responsible for approving documents and for reviewing them, how to identify the changes and revision status, how to distribute the

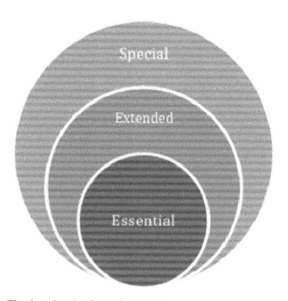

**FIGURE 2.3**    The three levels of security controls.

**TABLE 2.4**   Structure of ISO 27001

| Title | Controls | | |
|---|---|---|---|
| | Total | Essential | |
| | | Section (standard) | No. |
| Information Security Policy | 2 | **5.1.1** | 1 |
| Information Systems Acquisition, Development and Maintenance | 16 | **12.2.1** | 5 |
| | | **12.2.2** | |
| | | **12.2.3** | |
| | | **12.2.4** | |
| | | **12.6.1** | |
| Organization of Information Security | 11 | **6.1.3** | 1 |
| Information Security Incident Management | 5 | **13.2.1** | 3 |
| | | **13.2.2** | |
| | | **13.2.3** | |
| Business Continuity Management | 5 | **14.1.1** | 5 |
| | | **14.1.2** | |
| | | **14.1.3** | |
| | | **14.1.4** | |
| | | **14.1.5** | |
| Human Resources Security | 9 | **8.2.1** | 3 |
| | | **8.2.2** | |
| | | **8.2.3** | |
| Compliance | 10 | **15.1.2** | 3 |
| | | **15.1.3** | |
| | | **15.1.4** | |
| | **133** | | **21** |

documents, etc. In other words, this procedure should define how the organization's bloodstream (the flow of documents) will function.

*Second,* the *procedure for internal audits* defines responsibilities for planning and conducting audits, how audit results are reported, and how the records are maintained. This means that the main rules for conducting the audit must be set (Alfantookh, 2009; Kosutic, 2010, 2013; Saleh et al., 2007a, 2007b).

*Third,* the *procedure for corrective action* should define how any non-conformity and its causes are identified, how the necessary actions are defined and implemented, what records are taken, and how the review of the actions is performed. The purpose of this procedure is to define how each corrective action should eliminate the cause of the nonconformity (Alfantookh, 2009; Kosutic, 2010, 2013; Saleh et al., 2007a, 2007b).

*Fourth,* the *procedure for preventive action* aims at eliminating the cause of the non conformity so that it would not occur in the first place (Alfantookh, 2009; Kosutic, 2010, 2013; Saleh et al., 2007a, 2007b).

### 2.4.2 BS 7799

The British Standard 7799 (BS 7799) consists of two parts. The first part contains the best practices for information security management. The second part is the information security management system specification with guidance. BS 7799 introduced the PDCA, the Deming's quality assurance model (Figure 2.2). BS 7799 has 10 controls, which address key areas of information security management (Humphreys et al., 1998; BS, 2012). Those controls are as follows:

*Information Security Policy for the Organization:* This activity involves a thorough understanding of the organization's business goals and its dependency on information security. This entire exercise begins with the creation of an IT Security Policy. The policy should be implementable, easy to understand and must balance the level of protection with productivity. The policy should cover all the important areas like personnel, physical, procedural and technical.

*Creation of Information Security Infrastructure:* A management framework needs to be established to initiate, implement and control information security within the organization. This needs proper procedures for approval of the information security policy, assigning of the security roles and coordination of security across the organization.

*Asset Classification and Control:* One of the most laborious but essential tasks is to manage an inventory of all the IT assets, which could include information assets, software assets, physical assets or other similar services. These information assets need to be classified to indicate the degree of protection.

*Personnel Security:* Human errors, negligence and greed are responsible for most thefts, frauds or misuse of facilities. Various proactive measures should be taken to make personnel screening policies, confidentiality agreements, terms and conditions of employment, and information security education and training.

*Physical and Environmental Security:* Designing a secure physical environment to prevent unauthorized access, damage and interference to business premises and information is usually the beginning point of any security plan. This involves physical security perimeters, physical entry control, creating secure offices, rooms, facilities, providing physical access controls, providing protection devices to minimize risks ranging from fire to electromagnetic radiation, providing adequate protection to power supplies and data cables are some of the activities.

*Communications and Operations Management:* Proper documentation of procedures for the management and operation of all information processing facilities should be established. This includes detailed operating instructions and incident response procedures.

*Access Control.* Access to information and business processes should be controlled by the business and security requirements. This will include defining access control policy and rules, user access management, user registration, privilege management, user password use and management, review of user access rights, network access controls, enforcing paths from user terminals to computer, user authentication, node authentication, segregation of networks, network connection control, network routing control, operating system access control, user identification and authentication, use of system utilities, application access control, monitoring system access and use and ensuring information security when using mobile computing and tele-working facilities.

*System Development and Maintenance:* Security should ideally be built at the time of inception of a system. Hence security requirements should be identified and agreed prior to the development of information systems. There should be a defined policy on the use of such controls, which may involve encryption, digital signature, use of digital certificates, protection of cryptographic keys and standards to be used for maintenance of information security management.

*Business Continuity Management:* A business continuity management process should be designed, implemented and periodically tested to reduce the disruption caused by disasters and security failures. This begins by identifying all events that could cause interruptions to business processes and, depending on the risk assessment, preparation of a strategy plan.

*Compliance:* It is essential that strict adherence is observed to the provision of national and international IT law pertaining to Intellectual Property Rights (IPR), software copyrights, safeguarding of organizational records, data protection and privacy of personal information, prevention of misuse of information processing facilities, regulation of cryptographic controls and collection of evidence.

### 2.4.3   PCIDSS

The Payment Card Industry Data Security Standard (PCIDSS) is a worldwide information security standard defined by the Payment Card Industry Security Standards Council. PCIDSS is a set of policies and procedures intended to optimize the information security of credit, debit and cash card transactions and protect cardholders against misuse of their personal information (Bonner et al., 2013; Morse & Raval, 2008, 2011; Shaw, 2009).

However, the PCIDSS specifies and elaborates on six major objectives (Colum, 2009; Laredo, 2008; Morse & Raval, 2008, 2011;; Rowlingson & Winsborrow, 2006). *First,* a secure network must be maintained in which transactions can be conducted. This requirement involves the use of firewalls that are robust enough to be effective without causing undue inconvenience to cardholders or vendors. Specialized firewalls are available for wireless networks, which are highly vulnerable to eavesdropping and attacks by malicious hackers.

*Second*, cardholder information must be protected wherever it is stored. Repositories with vital data such as dates of birth, mothers' maiden names, social security numbers, phone numbers and mailing addresses should be secured against hacking.

*Third*, systems should be protected against the activities of malicious hackers by using frequently updated anti-virus software, anti-spyware programs, and other anti-malware solutions. All applications should be free of bugs and vulnerabilities that might open the door to exploits in which cardholder data could be stolen or altered.

*Fourth*, access to system information and operations should be restricted and controlled. Cardholders should not have to provide information to businesses unless those businesses must know that information to protect them and effectively carry out a transaction. Cardholder data should be protected physically as well as electronically.

*Fifth*, networks must be constantly monitored and regularly tested to ensure that all security measures and processes are in place, are functioning properly, and are kept up-do-date. For example, anti-virus and anti-spyware programs should be provided with the latest definitions and signatures.

*Sixth*, a formal information security policy must be defined, maintained, and followed at all times and by all participating entities. Enforcement measures such as audits and penalties for non-compliance may be necessary

Moreover, PCIDSS specifies the 12 requirements for compliance, which are called control objectives. Table 2.5 summarizes all those 12 control objectives.

### 2.4.4 INFORMATION TECHNOLOGY INFRASTRUCTURE LIBRARY

The Information Technology Infrastructure Library (ITIL) concept, emerged in the 1980s, it is a set of controls and practices for Information Technology Services Management (ITSM), Information Technology (IT) development and IT operations, which has some parts that focus on information security (Figure 2.4). ITIL describes procedures, tasks and checklists, used by an organization for establishing a level of competency. It allows the organization to establish a baseline from which it can plan,

**TABLE 2.5**   12 Controls Objectives

| Control Objectives | PCI DSS Requirements |
|---|---|
| Build and Maintain a Secure Network | 1. Install and maintain a firewall configuration to protect cardholder data |
| | 2. Do not use vendor-supplied defaults for system passwords and other security parameters |
| Protect Cardholder Data | 3. Protect stored cardholder data |
| | 4. Encrypt transmission of cardholder data across open, public networks |
| Maintain a Vulnerability Management Program | 5. Use and regularly update anti-virus software on all systems commonly affected by malware |
| | 6. Develop and maintain secure systems and applications |
| Implement Strong Access Control Measures. | 7. Restrict access to cardholder data by business need-to-know |
| | 8. Assign a unique ID to each person with computer access |
| Regularly Monitor and Test Networks | 9. Restrict physical access to cardholder data |
| | 10. Track and monitor all access to network resources and cardholder data |
| Maintain an Information Security Policy | 11. Regularly test security systems and processes |
| | 12. Maintain a policy that addresses information security |

implement and measure. It is used to demonstrate compliance and to measure improvement, and it consists of five core controls as follows (ITGI, 2008; Shahsavarani & Ji, 2014; Toleman, 2009):

*Service Strategy:* It provides guidance on clarification and prioritization of service-provider investments in services. More generally, Service Strategy focuses on helping IT organizations improve and develop over the long-term. In both cases, Service Strategy relies largely upon a market-driven approach. Key topics covered include service value definition, business-case development, service assets, market analysis, and service provider types. Below is a list of covered processes:

1.   Strategy Management
2.   Service Portfolio Management
3.   Financial management for IT services

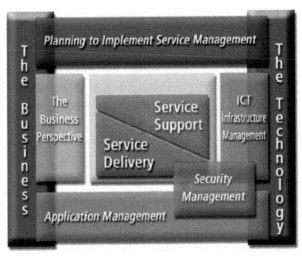

**FIGURE 2.4**   The three levels of security controls.

4. Demand Management
5. Business relationship management

*Service Design:* It provides good-practice guidance on the design of IT services, processes, and other aspects of the service management effort. Within ITIL, design work for an IT service is aggregated into a single service design package (SDP). Service design packages, along with other information about services, are managed within the service catalogs. Below is a list of the covered processes:

1. Design coordination
2. Service Catalogue
3. Service level Management
4. Availability Management
5. Capacity Management
6. IT Service Continuity Management (ITSCM)
7. Information Security Management System
8. Supplier Management

*Information Security Management System:* The ITIL-process Security Management describes the structured fitting of information security in the management of an organization. ITIL security management is based on the code of practice for ISO 27001.

The basic goal of security management is to ensure adequate information security. The primary goal of information security, in turn, is to protect information assets against risks, and thus to maintain their values to the organization. This is commonly expressed in terms of ensuring their confidentiality, integrity and availability, along with related properties or goals such as authenticity, accountability, non-repudiation and reliability.

*Service Transition:* It relates to the delivery of services required by a business into live/operational use, and often encompasses the "project" side of IT rather than business as usual (BAU). This area also covers topics such as managing changes to the BAU environment.

The list of ITIL processes in service transition:

1. Transition planning and support
2. Change management
3. Service asset and configuration management
4. Release and deployment management
5. Service validation and testing
6. Change evaluation
7. Knowledge management

*Service Operation.* It is aimed to provide the best practice for achieving the delivery of agreed levels of services both to end-users and the customers. Service operation is the part of the lifecycle where the services and value is actually directly delivered. The functions include technical management, application management, operations management and service desk as well as responsibilities for the staff engaging in service operation.

### 2.4.5 CONTROL OBJECTIVES FOR INFORMATION AND RELATED TECHNOLOGY

Control Objectives for Information and related Technology (COBIT) is an IT governance standard and supporting toolset that allows managers to bridge the gap between control requirements, technical issues, business risks, and security issues (Al Omari et al., 2012; De Haes et al., 2013; Oliver & Lainhart, 2012; Ridley et al., 2004). The COBIT Process Assessment Model (PAM) using COBIT 5 provides a basis for assessing an organization's processes. The model is evidence-based to enable a reliable, consistent and repeatable way to assess IT process capabilities, which helps IT leaders gain capability level and board member buy-in for change and improvement initiatives. Moreover, COBIT has five IT Governance areas of concentration (Al Omari et al., 2012; De Haes et al., 2013; Oliver & Lainhart, 2012; Ridley et al., 2004) (Figure 2.5):

- Strategic alignment, which focuses on ensuring the linkage of business and IT plans; defining, maintaining and validating the IT value proposition; and aligning IT operations with enterprise operations.

- Value delivery, which is about executing the value proposition throughout the delivery cycle, ensuring that IT delivers the promised benefits against the strategy, concentrating on optimizing costs and proving the intrinsic value of IT.

- Resource management, which is about the optimal investment and the proper management of critical IT resources: applications, information, infrastructure and people.

- Risk management, which is a clear understanding of the enterprise's appetite for risk, understanding of compliance requirements, and transparency into the organization.

- Performance measurement, which tracks and monitors strategy of implementation, project completion, resource usage, process performance and service delivery, for example, balanced scorecards that translate strategy into action to achieve goals measurable beyond conventional accounting.

- The business orientation of COBIT consists of linking business goals to IT goals, providing metrics and maturity models to measure their achievement, and identifying the associated responsibilities of business and IT process. The focus process of COBIT is illustrated by a process model that subdivides IT into four domains (Plan and Organize, Acquire and Implement, Deliver and Support, and Monitor and Evaluate) in line with the responsibility areas of plan, build, run and monitor. It is positioned at a high level and has been aligned and harmonized with others, such as IT standards and good practices such as ITIL, ISO 27001, The Open Group Architecture Framework

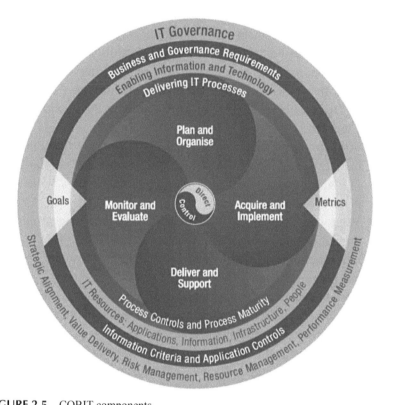

**FIGURE 2.5**   COBIT components.

(TOGAF) and Project Management Body of Knowledge (PMBOK). The COBIT components include Ridley et al., 2004; De Haes et al., 2013; Oliver & Lainhart, 2012; Al Omari et al., 2012:

- Framework: Organising IT governance objectives and good practices by IT domains and processes, and link them to business requirements.
- Process description: A reference process model and common language for everyone in an organization. The processes map to responsibility areas of planning, building, running and monitoring.
- Control objectives: Provide a complete set of high-level requirements to be considered by management for effective control of each IT process.

- Management guidelines: Help assign responsibility, agree on objectives, measure performance, and illustrate interrelationship with other processes.
- Maturity models: Assess maturity and capability per process and helps to address gaps.

### 2.4.6 STANDARDS: A SUMMARY OF FACTS

Alfantookh (2009), Bakry (2003b), Calder & Watkins (2012), Ridley et al. (2004); De Haes et al. (2013), have indicated mandatory requisition controls that should be implemented by an organization as critical requirements of the information security criteria, due to features as a basis of parameters for the fulfillment of standards. The details of those mandatory requisition controls are as follows:

1. *Information Security Policy:* how an institution expresses its intent with regards to information security, meaning by which an institution's governing body expresses its intent to secure information. Gives direction to management and staff and informs the other stakeholders of the primacy of efforts.
2. *Communications and Operations Management:* defines policy on security in the organization, in reducing security risk and ensuring correct computing, including operational procedures, controls, and well-defined responsibilities.
3. *Access Control:* is a system which enables an authority to control access to areas and resources in a given physical facility or computer-based information system.
4. *Information System Acquisition, Development and Maintenance:* an integrated process that defines boundaries and technical information systems, beginning with the acquisition, development and the maintenance of information systems.
5. *Organization of Information Security:* a structure owned by an organization in implementing information security, consisting of: the management's commitment to information security, information security co-ordination and authorization process for

information processing facilities. Two major directions: internal organization and external parties.

6. ***Asset Management:*** based on the idea that it is important to identify, track, classify, and assign ownership for the most important assets to ensure they are adequately protected.

7. ***Information Security Incident Management:*** a program that prepares for incidents. From a management perspective, it involves identification of resources needed for incident handling. Good incident management will also help with the prevention of future incidents.

8. ***Business Continuity Management:*** to ensure continuity of operations under abnormal conditions. Planning promotes the readiness of institutions for rapid recovery in the face of adverse events or conditions, minimizing the impact of such circumstances, and providing means to facilitate functioning during and after emergencies.

9. ***Human Resources Security:*** to ensure that all employees (including contractors and users of sensitive data) are qualified and understand their roles and responsibilities of their job duties and that access is removed once employment is terminated.

10. ***Physical and Environmental Security:*** the measures taken to protect systems, buildings, and related supporting infrastructure against threats associated with their physical environment. Buildings and rooms that house information and information technology systems must be afforded appropriate protection to avoid damage or unauthorized access to information and systems.

11. ***Compliance:*** these issues are necessarily divided into two areas; the first area involves compliance with the myriad laws, regulations or even contractual requirements which are part of the fabric of every institution. The second area is compliance with information security policies, standards and processes.

Moreover, each standard's overview and summary on their respective positions are shown in Table 2.6, and head-to-head comparisons between the big five ISMS standards deal with mandatory requisition controls of information security (Table 2.7) (Susanto et al., 2011a, 2011b, 2012a, 2012b).To help organizations manage their information security effectively, their information security systems have to comply with an ISMS standard. There are several such standards that lead to information security guidelines, namely PRINCE2, OPM3, CMMI, P-CMM, PMMM, ISO 27001, BS7799, PCIDSS, COSO, SOA, ITIL and COBIT. Some of these standards are not well-adopted by the organizations, for a variety of reasons. The big five ISMS standards are ISO 27001, BS 7799, PCIDSS, ITIL and COBIT. They are widely used standards for information security. The big five standards are discussed and compared by Susanto et al. (2012a) to determine their respective strengths, focuses, main components and adoption. In conclusion, each standard plays its own role, ISO 27001 and BS 7799 focusing on ISMS, while PCIDSS focuses on information security related to online business transactions and smart cards. ITIL and COBIT focus on project management and ICT Governance (Figure 2.6). If we refer to the usability of each standard then ISO 27001 is the leading standard. This indicates that ISO 27001 is well-adopted and well recognized by stakeholders (top management, staff, suppliers, customers/clients, regulators).

## 2.5 INFORMATION SECURITY MANAGEMENT SYSTEM FRAMEWORK

Since information security has a very important role in supporting activities for organizations, a framework that helps to comply with information security standards is needed (Calder & Watkins, 2010, 2012). Although the development of ICT security frameworks have gained momentum in recent years, more works on approaches to security framework are still needed. This section discusses several information security frameworks (Table 2.8) in the literature.

There are nine state-of-the-art frameworks (9STAF) available to help conceptualize and organize efforts to comply with information security standards. Those frameworks are: (1) a Framework for the Governance of Information Security (Posthumus and Solms, 2004); (2) a Framework for Information Security Management based on Guiding Standards: a United States perspective (Sipior and Ward, 2008); (3) a Security Framework for Information Systems Outsourcing (Fink, 1994); (4) Information Security Management: a Hierarchical Framework for Various Approaches (Eloff

**TABLE 2.6** Profile of Big Five of ISMS Standards

| | ISO 27001 | BS 7799 | PCIDSS | ITIL | COBIT |
|---|---|---|---|---|---|
| *Profile of Standards* | *ISO is a **non-governmental organization** that forms a bridge between the public and private sectors. On the one hand, many of its member institutes are part of the governmental structure of their countries, or are mandated by their government; also other members have their roots uniquely in the private sector, having been set up by national partnerships of industry associations* | *BS 7799 Standards is the UK's National Standards Body (NSB). BS Standards works with manufacturing and service industries, businesses, governments and consumers to facilitate the production of British, European and international standards* | *PCIDSS is a worldwide information security standard defined by the Payment Card Industry Security Standards Council. The standard was created to help industry organizations that process card payments prevent credit card fraud through increased controls around data and its exposure to compromise* | *ITIL is the abbreviation for the guideline IT Infrastructure Library, developed by CCTA, now the OGC (Office of Governance Commerce) in Norwich (England) developed on behalf of the British government. The main focus of the development was on mutual best practices for all British government data centers to ensure comparable services* | *COBIT is an IT governance framework and supporting toolset that allows managers to bridge the gap between control requirements, technical issues and business risks. COBIT enables clear policy development and good practice for IT control throughout organizations. COBIT emphasizes regulatory compliance, helps organizations to increase the value attained from IT* |
| *Initiated by* | *Delegates from 25 countries* | *United Kingdom Government's Department of Trade and Industry (DTI)* | ***VisaCard, MasterCard, American Express, Discover Information** and Compliance, and the **JCB** Data Security Program* | *The Central Computer and Telecommunications Agency (CCTA), now called the Office of Government Commerce (OGC)– UK* | *Information Systems Audit and Control Association (ISACA) and the IT Governance Institute (ITGI)–USA* |

| | ISO 27001 | BS 7799 | PCIDSS | ITIL | COBIT |
|---|---|---|---|---|---|
| **Launched on** | *February 23, 2005* | *1995* | *15 December 2004* | *1980s* | *1996* |
| **Standards & Components** | **18,500 International Standards** | **27,000 active standards** | **6 main components on standard** | **8 main components + 5 components version 3** | **6 main components on standard** |
| **Certificate Name** | *Certificate of ISO 27000 Series* | *Certificate of BS 7799: 1-2* | *Certificate of PCI-DSS Compliance* | *Certificate of ITIL Compliance* | *Certified Information Systems Auditor™ (**CISA®**)*; *Certified Information Security Manager® (**CISM®**)*; *Certified in the Governance of Enterprise IT® (**CGEIT®**)*; *Certified in Risk and Information Systems Control™ (**CRISCTM**)* |
| **Scope** | *Information Security* | *Information Security* | *Information and Data Transaction Security on debit, credit, prepaid, e-purse, ATM, and POS* | *Service Management* | *IT Governance* |
| **Usability** | *163 national members out of the 203 total countries in the world* | *110 national members out of the 203 total countries in the world* | *125 countries out of the 203 total countries in the world* | *50 international chapters* | *160 countries* |

**TABLE 2.7** Features Comparisons of the Big Five ISMS Standards

| S. No. | Standards | ISO 27001 | BS 7799 | PCID SS V2.0 | ITIL V4.0 | COBIT V4.1 |
|---|---|---|---|---|---|---|
| 1. | Information Security Policy | √ | √ | √ | √ | √ |
| 2. | Communications and Operations Management | √ | √ | √ | • | √ |
| 3. | Access Control | √ | √ | √ | √ | √ |
| 4. | Information Systems Acquisition, Development and Maintenance | √ | √ | √ | • | √ |
| 5. | Organization of Information Security | √ | √ | √ | √ | √ |
| 6. | Asset Management | √ | √ | • | √ | √ |
| 7. | Information Security Incident Management | √ | • | √ | √ | √ |
| 8. | Business Continuity Management | √ | √ | √ | √ | √ |
| 9. | Human Resources Security | √ | √ | √ | • | • |
| 10. | Physical and Environmental Security | √ | √ | √ | • | √ |
| 11. | Compliance | √ | √ | √ | √ | √ |

and Solms, 2000); (5) Information Security Governance Framework (Ohki et al., 2009); (6) Queensland Government Information Security Policy Framework (QGISPF, 2009); (7) STOPE methodology (Bakry, 2004); (8) a Security Audit Framework for Security Management in the Enterprise (Onwubiko, 2009); (9) Multimedia Information Security Architecture Framework (Susanto & Muhaya, 2010). Table 2.1 shows briefly the profile and features of the frameworks.

The Framework for the Governance of Information Security (FGIS) was introduced by Posthumus and Solms (2004), which reveals and suggests the important part of protecting an organization's vital business information assets. This action should be considered as such and should be included as a part of corporate governance responsibility by the corporate executives.

Sipior and Ward (2008) introduced the second framework which intends to promote a cohesive approach, which considers a process view of information within the context of the entire organizational operational environment.

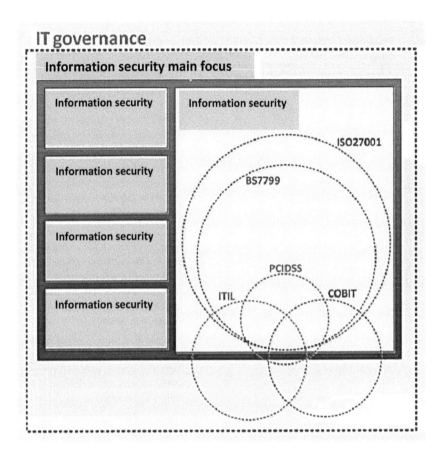

**FIGURE 2.6**   Position of the big five standards.

Fink (1994) suggested the Security Framework for Information Systems Outsourcing, which assumes that most of information systems (IS) related matters can be outsourced. Consequently, information security issues are controlled by the outsourced company. The aim of this framework is to evaluate the loss in IS security and control for IS outsourcing.

The Hierarchical Framework for Information Security Management has a principal aim of assisting management in the interpretation as well as in the application of internationally accepted approaches to IS management (Eloff & Solms, 2000).

**TABLE 2.8**   Existing Frameworks for Information Security

| S. No. | Existing information security frameworks | Authors | Framework Focus |
|---|---|---|---|
| 1. | A framework for the governance of information security | Posthumus and Solms, 2004 | • The importance of protecting an organisation's vital business information assets by investigating several fundamental considerations of corporate executives' concern. |
| 2. | A Framework for Information Security Management Based on Guiding Standard: A United States Perspective | Sipior and Ward, 2008 | • Intended to promote a cohesive approach that considers a process view of information within the context of the entire organisational operational environment. |
|  |  |  | • The four levels of information security: international oversight of information security, national oversight of information security, organizational oversight of information security, employee oversight of information security. |
| 3. | A Security Framework for Information Systems Outsourcing | Fink, 1994 | • Assumes that most of information security related matter can be outsourced. Consequently, information security issues are controlled by the outsourced company. |
|  |  |  | • The aim of this framework is to evaluate the loss in information security and control when information security outsourcing occurs. |
| 4. | Information Security Management: a Hierarchical Framework for Various Approaches | Eloff and Solms, 2000 | • Introduces and elucidates ill-defined terms and concepts of information security. |
|  |  |  | • All organizations are dependent on their information security resources in today's highly competitive global markets, not only for their survival but also for their growth and expansion. |
| 5. | Information Security GovernanceFramework | Ohki et al., 2009 | • Supports corporate executives to direct, monitor, and evaluate ISMS related activities in a unified manner |

**TABLE 2.8** (Continued)

| S. No. | Existing information security frameworks | Authors | Framework Focus |
|---|---|---|---|
| 6. | Queensland Government Information Security Policy Framework | QGISPF, 2009 | • Identifies various areas which contribute to effective information management and serves as an organizing framework for ensuring appropriate policy coverage and avoiding overlaps which may occur without such a framework. |
| 7. | STOPE Methodology | Bakry, 2004 | • Introduces domains of attention in information security to better focus the analysis of existing problems from the perspectives of Strategy, Technology, Organisation, People and Environment. |
| 8. | A Security Audit Framework for Security Management in the Enterprise | Onwubiko, 2009 | • Comprises five components and three subcomponents. The components consist of security policies that define acceptable use, technical controls, management standards and practices. |
| | | | • Stipulates acceptable regulatory and security compliances. |
| 9. | Multimedia Information Security Architecture Framework | Susanto and Muhaya, 2010 | • Emphasises on information security in multimedia objects. Proposed the Multimedia Information Security Architecture which the authors based on ISA architecture which has 5 main components, namely: Security Policy, Security Culture, Monitoring Compliance, Security Program and Security Infrastructure. |
| | | | • MISA consist of 8 main components, namely: *Security Governance, Security & Privacy, Multimedia Information Sharing, Protection & Access, Security Assessment, Cyber Security, Enterprise Security, and Security Awareness.* |

Ohki et al. (2009) introduced the Information Security Governance (ISG) framework, which combines and inter-relates existing information security schemes; Corporate Executives can direct, monitor, and evaluate ISMS related activities in a unified manner, in which all critical elements are explicitly defined with the functions of each element and interfaces among elements.

The Government Information Security Policy Framework (GISPF) defines the generic classification scheme for information security policies, and does so with a perspective that it is independent from the physical implementation models chosen by departments, agencies, and offices. GISPF identifies and defines various areas which contribute to effective information management and serve as an organizing framework for ensuring appropriate policy coverage and avoiding overlaps that may occur (QGISPF, 2009).

The Security Audit Framework for Security Management in the Enterprise comprises five components and three subcomponents. The components comprise security policies that define acceptable use, technical controls, management standards, and practices. It focuses not only on technical controls around information security, but also processes, procedures, practice and regulatory compliance that assist organizations in maintaining and sustaining consistently high quality information security assurance (Onwubiko, 2009).

Susanto and Muhaya (2010) introduced the Multimedia Information Security Architecture Framework which emphasizes on information security in multimedia objects and issues in existing architecture. Multimedia Information Security Architecture (MISA) authors proposed the framework based on the 5 components of ISA architecture. MISA architecture itself has 8 major components (Table 2.8).

STOPE methodology introduces domains of attention in information security to better focus the analysis of existing problems from the perspectives of strategy, technology, organization, people and environment. These elements are identified in the following explanation. *Strategy:* the strategy of the country with regards to the future development of the industry or the service concerned. *Technology:* the technology upon which the industry or the service concerned is based. *Organization:* the organizations associated with or related to, the industry or the service concerned. *People:* the people concerned with the target industry or service. *Environment:* the environment surrounding the target industry or service (Bakry, 2004).

## 2.6 EXISTING METHODS AND TOOLS TO AID IN ADOPTING ISO 27001

The above frameworks address general approaches or methods to handle information security. This section discusses specific approaches and methods to help organization comply with ISO 27001.

Gillie (2011) proposed an approach called the five stages to information security (5S2IS). 5S2IS uses five major steps in assessing the readiness of an organization to comply with ISO 27001. Those steps are *draw up a plan, define protocols, organization measurement, monitoring*, and *improvement* (the details are explained in the Subsection 2.6.1).

Montesino (2012) conducted an analysis and pilot implementation of ISO 27001 using security information and event management (SIEM). This system works by analyzing the computing system, either through log data or through real time. SIEM consists of security information management (SIM) and security event management (SEM) (the details are explained in the Subsection 2.6.2).

Kosutic (2010, 2013) described the use of another method known as the implementation checklist method (ICM). This method pursues the parameters of the standard. The main purposes of ICM are: (1) to gauge the level of compliance to ISO 27001 requirements by holding, department, and division; (2) facilitate the provision of information necessary for ISO 27001 implementation; and (3) serve as training materials for understanding the ISO 27001 requirements (The details are explained in the Subsection 2.6.3).

All those features along with the methods and tools are shown in Table 2.9. For instance, 5S2IS is the abstraction methodology without any implementing and practical tools such as software to assist organizations to perform self-assessment. On the other hand, SIEM can be used to capture network parameters. Note that the stability of the network and security is a key element in obtaining ISO 27001 compliance. The limitation of SIEM is that non-technical parameters, such as strategic and managerial parameters are not covered. ICM describes 133 controls and users are given a checklist as a medium to control which parts have been implemented and which parts have not. ICM is available in electronic format (an excel file). The user checks manually one-by-one, the controls chosen then explain how and why they are appropriate, which are applicable and which are not, and the reasons for such a decision.

### 2.6.1   FIVE STAGES TO INFORMATION SECURITY (5S2IS)

Gillies (2011) developed 5S2IS to encourage an organization's compliance with ISO 27001. 5S2IS is built on the foundations of ISO 27001, and the Capability Maturity Model (CMM). The approach shows any defects and improves quality, feeding information into an internal ISO assessment. Schematically, the approach maps the plan-do-check-act cycle onto a five-stage development process, shown in Figure 2.7.

The key to the 5S2IS approach is to draw together these disparate elements to systematize the approach. Each stage of the process reduces risk and further protects the information against risks:

- Stage 1. Draw up a plan. This stage is about defining smart goals for each of the dimensions to make explicit what the organization is seeking to achieve. As well as defining the goals, an organization will need to get its management to sign up to the goals, to the process to be followed to reach these goals, and to the target stage to be reached and whether or not the ultimate goal is to demonstrate compliance with ISO 27001 through external certification.
- Stage 2. Define a protocol for each of the smart goals defined in Stage 1. The protocol should define what the organization will do in order to achieve each goal. It will define the processes and outline compliance measures that will be put in place to demonstrate whether the organization is following the protocols.
- Stage 3. Measure the organization's performance against the protocols from Stage 2. In particular, identify non-compliances with the protocols. This stage represents the implementation of the defined protocols. It will require the collection of monitoring data as defined in Stage 2. It is also associated with significant

**TABLE 2.9** Comparisons of Existing Assessment Tools

| Features and Function | Measurement tool | Monitoring tool | Controls | Note |
|---|---|---|---|---|
| 5S2IS (Gillies, 2011) | • Maps the plan-do-check-act cycle onto a five-stage development process of adoption. | • 5S2IS is a methodology proposed to help an organisation implement ISO 27001<br>• It is also supported by CMM (capability maturity model). No (computer) tool is provided with this method to measure an organisation's compliance level. | Covered 11 controls | Lack of practical implementation (tool or software) |
| SIEM (Montesino, 2012) | • Analyses security event data in real time;<br>• Collects, stores, analyses and Analyses reports on log data for regulatory compliance and forensics. | – • Security information management (SIM).<br>• Provides the collection, reporting and analysis of log data.<br>• Supports regulatory compliance reporting, internal threat management and resource access monitoring. | – • Security event management (SEM).<br>• Real-time monitoring and incident management for security-related events.<br>• Processes log and event data from security devices, network devices, systems and applications in real time<br>• Provide security monitoring, event correlation and incident esponses. | Covered 10 controls | Lack of RISC investigation that covered essential controls of ISO 27001 |

**TABLE 2.9** (Continued)

| Features and Function | Measurement tool | Monitoring tool | Controls | Note |
|---|---|---|---|---|
| • ICM (Kosutic, 2010, 2013) | • Provides a method to comply with the parameters of the standard<br>• Helps gauge the level of compliance to ISO 27001<br>• Facilitate the provision of information necessary for ISO 27001 implementation. | • It provides a checklist sheet associated with ISO 27001 controls<br>• Helps stakeholders to review and check their information security condition through ICM<br>• However, no specific tools were mentioned to measure and assess an organisation's readinesslevel | Covered 133 controls | Lack of electronically checking for controls chosen, appropriate explanation, and the reasons for such a decision. |
| ISM | • RISC Investigation<br>• Real-time network monitoring<br>• Organisations self-assess<br>• Knowledge and information for ISO 27001 are available in the system. | • E-assessment through RISC investigation module<br>• Strengths-weaknesses analysis<br>• Level by level investigation result<br>• e-monitoring through SMM features<br>• Real-time monitoring<br>• Firewall management<br>• Process information<br>• Port detection | Covers 21 essential controls, and can be expanded to 144 controls | Provides electronic checking for ISO 27001 essential controls (SAM), and also a monitoring system for suspected information security breaches (SMM) in an integrated software solution. |

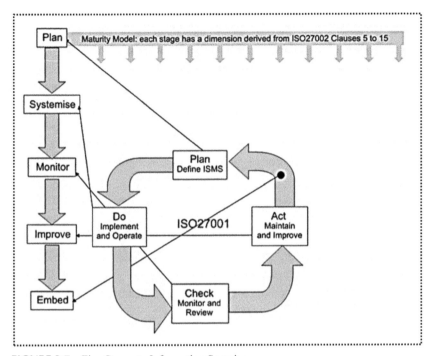

**FIGURE 2.7**    Five Stages to Information Security.

cultural change within the organization as the protocols move from strategic commitment and then definition into implementation. It requires significant acceptance and ownership from the staff across the organization.

- Stage 4. Use the monitoring data from Stage 3 to improve performance and reduce non-compliances. This stage will require root cause analysis to identify underlying problems rather than superficial symptoms.
- Stage 5. Embed the improvement cycle within the organization. At this stage the organization should have an ISMS compatible with ISO 27001, and they may choose to move forward to ISO 27001 certification for verification, credibility and marketing purposes.

## 2.6.2   SECURITY INFORMATION AND EVENT MANAGEMENT (SIEM)

Montesino (2012), and Nicolett and Kavanagh, (2011) described the SIEM technology as a function to analyze its possibilities for automation and integration of security tools as a requirement of ISO 27001. SIEM technology is used to analyze security event data in real time; and to collect, store, analyze and report on log data for regulatory compliance and forensics. It is composed of two main functions (Nicolett and Kavanagh, 2011):

1. Security information management (SIM). Log management and compliance reporting. SIM provides the collection, reporting and analysis of log data from network devices. This data functions to support regulatory compliance, internal threat management and resource access monitoring.

2. Security event management (SEM). Real-time monitoring and incident management for security-related events. SEM processes create logs and event data from security devices, network devices, systems and applications in real time to provide security monitoring, event correlation and incident responses. SIEM tools collect event data in near real time in a way that enables immediate analysis. Data collection methods include:

   ▪ A syslog data stream from the monitored event source;
   ▪ agents installed directly on the monitored device;

Despite the evolution of SIEM, the CSI survey (*Annual Computer Crime and Security Survey*, Computer Security Institute) does not consider SIEM solutions among the security technologies deployed to protect organizations, and only 46.2% of respondents use log management software (Richardson, 2011). On the other hand, the SANS log management survey (Sixth Annual Log Management Survey Report) shows that the top three reasons for collecting logs and using SIEM systems were: detecting incidents, determining what happened (forensics and analysis), and meeting compliance requirements; unfortunately less than 20% of respondents believe that SIEM systems can support other IT operations and reduce cost for IT security.

## 2.6.3   IMPLEMENTATION CHECKLIST METHOD (ICM)

An Implementation checklist method (ICM) pursues the parameters of the standard using a checklist. The purposes of its implementation are: (1) to gauge the level of compliance to ISO 27001 requirements by holding, department, and division. (2) Facilitate the provision of information necessary for ISO 27001 implementation. (3) Serve as training materials for understanding the ISO 27001 requirements. Moreover, implementation of the checklist consists of 16 steps as follows (Kosutic, 2010, 2013):

*Obtain Management Support,* this stage may seem rather obvious but it is the main reason why ISO 27001 projects fail, since management does not provide enough people to work on the project or not enough budget to deal with it (see Subsection 2.3.1.1).

*Treat it as a Project,* an ISO 27001 implementation is a complex issue involving various activities, lots of people, lasting several months (or more than a year). An organization must define clearly what is to be done, who is going to do it and in what time frame (i.e., apply project management).

*Define the Scope,* if the stakeholder is a large organization, it probably makes sense to implement ISO 27001 only in one part of an organization, to limit and localize the issues of compliance (e.g., only the ICT department would comply with ISO 27001), then after its success, the stakeholder could expand to other departments.

*Write an ISMS Policy,* ISMS Policy is the highest-level document in ISMS – it should not be very detailed, but it should define some basic issues for information security. The purpose is for management to define what it wants to achieve, and how to control it.

*Define the Risk Assessment Methodology,* risk assessment is the most complex task in the ISO 27001 project – the point is to define the rules for identifying the assets, vulnerabilities, threats, impacts, and to define the acceptable level of risk.

*Perform the risk assessment & risk treatment,* here the stakeholder has to implement what were defined in the previous step. The point is to get a comprehensive picture of the vulnerabilities of the organization's information.

*Write the Statement of Applicability (SoA)*, once the risk treatment process is completed, the stakeholder will know exactly which controls are needed (there are a total of 133 controls but the stakeholder might not need them all). The purpose of SoA is to list all controls and to define which are applicable and which are not, and the reasons for such a decision, the objectives to be achieved with the controls and a description of implementation.

*Write the Risk Treatment Plan*, to define how the controls from SoA are to be implemented – who is going to do it, when, with what budget, etc. This document is actually an implementation plan focused on the selected controls.

*Define How to Measure the Effectiveness of Controls*, to define how the stakeholder is going to measure the fulfillment of objectives it has set both for the whole ISMS, and for each applicable control in the SoA.

*Implement the Controls and Mandatory Procedures*, it means the application of new technology, but above all – implementation of new behavior in an organization. Often new policies and procedures are needed (meaning that change is needed), and people usually resist change – this is why the next task (training and awareness) is crucial for avoiding that risk.

*Implement Training and Awareness Programs*, to implement all the new policies and procedures, the stakeholder has to explain to personnel why it is necessary, and train them to be able to perform as expected.

*Operate the ISMS*, this is the part where ISO 27001 becomes an everyday routine in an organization; the crucial point here is keeping records. Using them, the stakeholder can monitor what is happening and will actually know with certainty whether employees (and suppliers) are performing their tasks as required.

*Monitor the ISMS*, this is where the objectives for controls and measurement methodology come together, checking whether the results are achieving the objectives. If not, the stakeholder has to perform corrective and/or preventive actions.

*Internal audit*, which functions to take corrective and preventive actions within an organization.

*Management Review*, which is to know what is going on in the ISMS, i.e., if everyone has performed his or her duties, if the ISMS is achieving desired results, etc.

*Corrective and Preventive Actions,* to identify the root cause of non-conformity, and then resolve, verify, correct, and prevent.

## 2.7 SOFTWARE DEVELOPMENT METHODOLOGIES AND PERFORMANCE MEASUREMENT

A software development process, also known as a software development life-cycle (SDLC), is a structure imposed on the development of a software product. Similar terms include *software life cycle* and *software process.* It is often considered a subset of systems development life cycle. There are several models for such processes, each describing approaches to a variety of tasks or activities that take place during the process. Some people consider a life-cycle model a more general term and a software development process as a more specific term. For example, there are many specific software development processes that 'fit' the spiral life-cycle model. ISO 12207 is an international standard for software life-cycle processes. It aims to be the standard that defines all the tasks required for developing and maintaining software (Ralph & Wand, 2009).

A decades-long goal has been to find processes that improve productivity and quality. Some try to systematize or formalize the seemingly unruly task of writing software. Others apply project management techniques to writing software. Without effective project management, software projects can easily be delivered late or over budget. With large numbers of software projects not meeting their expectations in terms of functionality, cost, or delivery schedule, effective project management for software development is definitely needed (Whitten et al., 2003). Further, organizations may create a Software Engineering Process Group (SEPG) for process improvement. SEPG is composed of line practitioners who have varied skills, the group is at the center of the collaborative effort of everyone in the organization who is involved with software engineering process improvement.

We combined several approaches of software development process to develop our system based on proposed framework. These approaches such as spiral development approach (SDA), waterfall approach (WFA), and release management approach (RMA) (Figure 2.8).

### 2.7.1 SPIRAL DEVELOPMENT APPROACH (SDA)

SDA is a software development process combining elements of both design and prototyping-in-stages, in an effort to combine advantages of top–down and bottom–up concepts (Figure 2.9). The basic principles of spiral development are (Boehm and Hansen, 2000, 2001):

* Focus on risk assessment to minimize project risk by breaking a project into smaller segments and providing more ease-of-change during the development process, as well as providing the opportunity to evaluate risks and weigh consideration of project continuation throughout the life cycle.

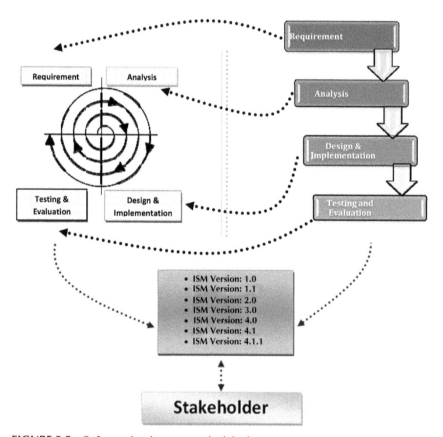

**FIGURE 2.8**　Software development methodologies.

- Each cycle involves a progression through the same sequence of steps, for each part of the product and for each of its levels of elaboration, from an overall concept-of-operation document down to the coding of each individual program.
- Each trip around the spiral traverses four basic quadrants: (1) analysis; (2) requirement; evaluate alternatives; identify and resolve risks; (3) design, implementation and verify deliverables from the iteration; and (4) testing, evaluation, and plan the next iteration.

### 2.7.2   WATERFALL APPROACH (WFA)

The WFA is a sequential design process, often used in software development processes, in which progress is seen as flowing steadily downwards (waterfall) through the phases of analysis, requirement, design,

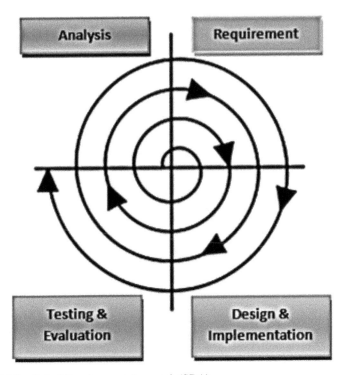

**FIGURE 2.9**   Spiral Development Approach (SDA).

construction, implementation, and testing (Cusumano and Smith, 1995; Huo et al., *I2004*) (Figure 2.10). WFA views the optimal process for software development as a linear or sequential series of phases that take developers from initial high-level requirements to system testing and evaluation. Designers begin by trying to write a specification as complete as possible. Next, they divide the specification into pieces or modules in a more detailed design phase.

The alternatives of WFA were proposed in the form of more iterative approaches to software development. These include notions of iterative enhancement as well as the spiral model of software development. These alternatives see developers moving around in phases, going back and forth between designing, coding, and testing as they move forward in a project. Because most software projects require extensive iterations among specification, development, and testing activities, developers have been trying to build software around reusable modules or objects, which means that developers need to begin with a detailed overview of the system objectives. This type of activity requires a departure from the linear waterfall steps, although many

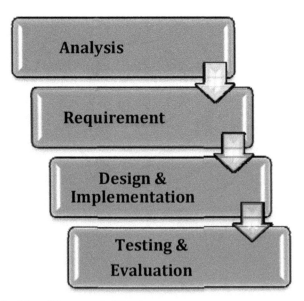

FIGURE 2.10   Waterfall approach (WFA).

companies known for pushing reusability tend to end with a single integration phase at the end of the project (Cusumano and Smith, 1995; MacCormack & Verganti, 2003).

## 2.7.3 RELEASE MANAGEMENT APPROACH (RMA)

RMA is used to control distribution and testing of software, including evaluation across the entire organization's ICT and infrastructure. Proper software control ensures the availability of the latest version, which functions as intended when introduced into existing ICT systems (ITGI, 2008). The goals of release management include:

- Planning the rollout of software,
- Designing and implementing procedures for the distribution and installation in ICT systems,
- Effectively communicating and managing expectations of the customer during the planning and rollout of releases,
- Controlling the distribution and installation in ICT systems.

An RMA consists of the new or changed software and/or hardware required to implement. Release categories include (ITGI, 2008; Müller et al., 2006):

- Major software releases and major hardware upgrades normally containing large amounts of new functionality, some of which may make fixes for redundant problems.
- Minor software releases and hardware upgrades, normally containing small enhancements and fixes, some of which may have already been issued as emergency fixes.
- Emergency software and hardware fixes, normally containing the corrections to a small number of known problems.

Releases can be divided based on the release unit into: (1) *delta release:* a release of only that part of the software which has been changed. For example, security patches. (2) *Full release:* the entire software program is deployed—for example, a new version of an existing application. (3) *Packaged release:* a combination of many changes—for example, an operating system image which also contains specific applications.

## 2.7.4   PERFORMANCE MEASUREMENT

Wahono (2006) mentioned that the quality of the software (software quality, SQ) or performance of the software (software performance, SP) is the theme of study and research in the history of the science of heredity software engineering. It began on what will be measured (whether process or product), how to measure the software, the measuring point of view and how to determine the parameters of software quality measurement. However, measuring the quality of software is not a cushy stage, when someone gives a very good review for software, others may not necessarily say the same thing. One's view point may be oriented to one side of the problem (e.g., on the reliability and efficiency of software), while others claimed that the software was bad from the point of view of another (such as reusability aspects of the design) (Wahono, 2006).

The first question that arises when discussing the measurement of SQ/SP is exactly what aspects are to be measured. SQ/SP can be viewed from the perspective of the software development process and the results of the product. And the final assessment is necessarily oriented to how the software can be developed as expected by the user. This departs from the notion of quality according to IEEE standard glossary of software engineering technology (IEEE, 1990) that states: *The degree to which a system, component, process meets customer or user needs or expectations.*

From the product point of view, measuring SQ/SP can use the standard of ISO 9126 or best practices developed by practitioners and software developers. McCall proposed taxonomy as the best practice of software measurement (McCall, 1977a, 1977b). Boehm et al., (1976), Cavano & McCall (1978), Jung et al., (2004) and Kan (2002) stated that a series of parameters as performance indicators were defined, such as: advantages of the software being dealt with, accuracy, precision, stability, ease of operation, user interfaces or graphical user interfaces. Therefore, in this study, the authors adopted, refined, and improved mentioned SQ/SP parameters to become eight fundamental parameters (8FPs) to measure performance and features of the framework (ISF) and software (ISM) (Boehm et al., 1976; Cavano & McCall, 1978; Jung et al., 2004; Kan, 2002; McCall, 1977a, 1977b) (Table 2.10).

All 8FPs reflect ISF-ISM performance, behavior, function, and back-office programming to handle the workload of self assessment and monitoring. The details and description of adoption parameters as ISF-ISM performance indicators are as follows:

1. ISM functions as information security self-assessment;
2. The benefit of ISF-ISM in helping the organization understand the ISMS standard's (ISO 27001) controls;
3. The extent to which the ISM can be used to understand information security standard terms and concepts;
4. ISM features;
5. ISM graphical user interface and user friendliness;
6. Analysis precision produced by ISM;

**TABLE 2.10**  Performance Parameters

| References | Parameters | Parameters to Measure and Evaluate the ISF-ISM | Scale |
|---|---|---|---|
| Woodside (2007); Gan (2006); Lucas (1971) | 1. Benefits and advantages of the software/tool | The benefits of the ISM in helping the organisation understand the ISMS standard | Measurement and evaluation on a scale of 0-4; |
| | 2. Accuracy | Final result precision produced by ISF-ISM | 0: not recommended (not implemented); |
| | 3. Precision | Analysis precision produced by ISF-ISM | 1: partially recommended (below average); |
| | 4. Stability | ISM Performance | 2: recommended (average); |
| | 5. Ease of operation | ISM features | 3: highly recommended (above average); |
| | 6. Graphical user interface | ISF-ISM graphical user interface and user friendliness | 4: excellent |
| Solms (2002); Bakry (2007) | 7. Understanding information security standard concepts | How the ISF-ISM can be used to understand information security standard terms and concepts | |
| | 8. Assessment issues | ISF-ISM functions as information security self-assessment | |

7.  Final result precision produced by ISM;
8.  ISM performance.

It is important for SQ/SP parameters to be measured quantitatively, in the form of numbers or grades that are easily understood by users. Consequently, it is necessary to set measurement for the parameters or attributes. According to McCall taxonomy (McCall, 1977a, 1977b), attributes are hierarchically structured, where the upper-level (high-level attributes) are called factors, and the lower level (low-level attributes) are called criteria. The factors indicate product quality attributes from the perspective of the user, while the criteria are product quality parameters from the perspective of the software. Factors and criteria have a causal relationship (cause-effect) (Peters & Pedrycz, 1998; Van Vliet et al., 1993).

SQ/SP evaluation of ISF-ISM following the 8FPs are measured by summation of all the criteria in accordance with a weighting factor which has been established (Bowen et al., 1985). Measurement formula is explained as follows:

$$F_a = \sum_{i=1}^{n} w_i c_i$$

$$F_a = \frac{w_1 c_1 + w_2 c_2 + w_3 c_3 + \ldots + w_n c_n}{n} \qquad (1)$$

where $F_a$ is the total value of factor $a$; $w_i$ is the weight for criterion $I$; $c_i$ is the value for criterion $i$.

Stages that must be followed in the measurement are:

· *Stage 1:* Define the criteria to measure a factor;
· *Stage 2:* Determine the weight (w) of each criterion (e.g., →1; if all criteria have the same level of effect toward the whole performance of the software);
· *Stage 3:* Determine the scale of the value criteria (e.g., → 0 <= C <= 4);
· *Stage 4:* Give the measurement value of each criterion;
· *Stage 5:* Calculate the total value of the formula [1].

## KEYWORDS

- ISF-ISM performance
- spiral development approach
- SQ/SP evaluation
- waterfall approach

# CHAPTER 3

# METHODOLOGY

## CONTENTS

### 3.1   INTRODUCTION

Research Methodology (RM) in common parlance refers to a search for knowledge. It can also be defined as a scientific and systematic search for pertinent information on a specific topic. However, a methodology does not set out to provide solutions but to offer the theoretical underpinning for understanding which method, set of methods or *"best practices"* can be applied to a specific case. It normally encompasses concepts such as paradigms, theoretical models, phases and quantitative or qualitative techniques employed in doing research (Blessing et al., 1998; Kothari, 2004; Scandura & Williams, 2000).

This chapter focuses on one of the key steps taken towards research success that is the adoption of RM. It begins with a description of the methodology as the foundation of study. In the context of this research, methodology discusses the procedures, objectives, respondents, software development issues, and data analysis about information security aspects.

This chapter covers four major components: research design, respondent types, fieldwork (data collection), and ISM testing-evaluation. There are four stages conducted in this research. The first stage is knowledge discovery through literature analysis on related work, comparative studies and information security standard (ISO 27001) refinement. Refinement of ISO 27001 is a deterministic process to determine the degree of clarity of each essential control over its parameters (Susanto et al., 2011a, 2011b). The second stage is to construct a new framework. The third stage is to develop software as a tool to conduct ISO 27001 investigation for compliance and RISC measurement. Finally, the fourth stage is conducting RISC investigations for respondent organizations to measure their compliance and provide comprehensive software evaluation to measure the software quality (SQ) and software performance (SP) by following McCall taxonomy (1977).

## 3.2 AN OVERVIEW OF METHODOLOGY STAGES

According to Von Solmn (2005a, 2005b) information security aspects have a very important role in supporting and enabling activities of the organization. Standish Group (2013) stated that many US ICT projects, including ISMS standardizing and ISO 27001 compliance in major organizations faced difficulties; many had reported failure and lost billions of dollars. BCS Review (2001) found that only around one in eight (13%) ICT information security-standardizing projects were successful. Kosutic (2010, 2013), AbuSaad et al., (2011), Mataracioglu & Ozkan (2011), Alshitri & Abanumy (2014) stated that technical barriers, the project owner's lack of understanding of processes, technical aspects, lack of internal ownership, and neglect of certain aspects, were the major problems that caused the delay for ISMS and ISO 27001 projects. One of the key components to aid in understanding the processes and technical aspects is by using a framework. Unfortunately, existing frameworks do not provide formal and practical models for RISC measurements[1]. This research proposes a framework that could grab those issues; formal and practical models for RISC measurement is needed. This framework is designed in such a way

---

[1] The RISC measurement is investigation stages to find out organization's readiness and information security capabilities over the ISO 27001.

to provide an integrated solution to overcome an organization's technical barriers and difficulties in understanding, investigating and compliance with the ISMS standard (ISO 27001). The proposed framework answers the technical aspects of the research questions such as the main barriers in implementing ISMS within an organization, the gaps between state-of-the-art existing frameworks and solutions to formal and quantitative investigation of RISC parameters. The proposed framework approach will influence an organization's learning and preparation time to comply with ISO 27001 and provide advantages for an organization through self-assessment by ISM for RISC measurement for ISO 27001 certification.

The RM in this study describes and explains the components and scenarios of the proposed framework and further system development. The research activities are summarized in Figure 3.1. The study combines FGC (focus group discussion), SDM (software development methodologies), RISC (readiness and information security capabilities) and SP/SQ (software performance and software quality) as quantitative measurement.

This roadmap will identify the sizes of respondent organizations, procedures and techniques to analyze and derive the data to construct the proposed framework and modeling software of ISM, and finally, test the software onsite to measure the outcome. The stages of methodology are literature analysis, proposed framework based on literature study, constructing a survey instrument to confirm the framework and grab user requirements and developing a full version software following SMD particularly WSP-SM (waterfall software process-spiral model development) of software development guidance.

### 3.2.1 STAGE 1: LITERATURE SURVEY

To develop a framework, this study is built upon recent reviews of frameworks and tools that have been developed. The relevant journal search was conducted using keywords indicating the area of the research, such as information security, security awareness, information security for marketing tools, information security standards, ISO 27001 implementation, information security framework, information security breaches, information security capabilities, e-assessment and e-monitoring for ISO 27001 compliance, essential controls, and ISO 27001 refinement. English

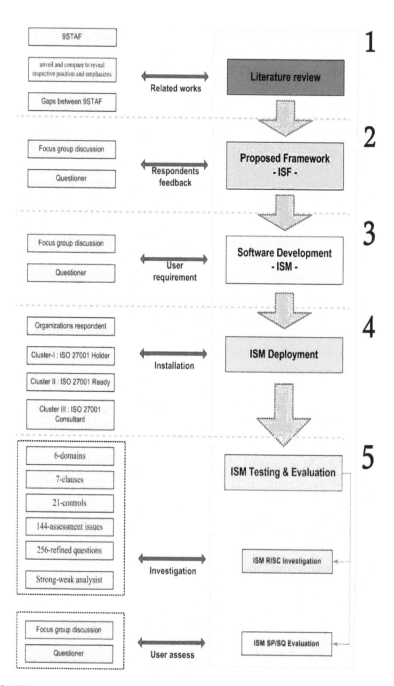

**FIGURE 3.1**    Research Stages.

peer-reviewed journals were chosen for further review and investigation. Since information security is a very interesting and challenging area, we successfully grabbed more than 200 scholarly journals that have strong relations with this study. We conducted a thematic analysis among articles and papers with mentioned keywords and area.

This research starts by conducting a literature survey to observe points of concern, advantages and disadvantages of the existing frameworks (the aforementioned 9STAF) (Figure 3.1, layer 1). The 9STAF are widely used in relation to information security issues. The 9STAF were studied carefully and a comparison among them was carried out to reveal their salient features and respective positions. The details of 9STAF's strengths and weaknesses are discussed further in Chapter 4, proposed framework. The objective to this is to determine the gaps and to establish the direction of this research. It is important to make sure that the new framework to be developed must overcome the weaknesses of the existing frameworks. For instance, none of the existing frameworks offer a method to measure RISC.

### 3.2.2 STAGE 2: PROPOSED FRAMEWORK

Once the proposed framework is developed, feedback from potential users is needed to make sure the framework will assist them in solving their problems. The potential users' feedback will be used to improve the framework and to determine user requirements for developing the software as a tool that can help them unveil their positions and at the same time measure the stage of RISC for compliance with ISO 27001. Note that the potential users should be those who are actively involved in various stages of ISO 27001 compliance projects.

The initial stage of ISF was developed based on literature review, especially the in-depth study of 9STAF. User feedback is used to improve and refine the ISF. The next step is to map components of ISO 27001 (such as control, clause, and assessment issues) to the framework. Each domain will be tested and measured by the respondents through the tool's user interface (more explanation in Chapter 5: Integrated Solution Modeling) to produce the investigation's indicator result. An indicator result is a score or grade that describes the organization's circumstances associated with the RISC level regarding ISO 27001 (Figure 3.1, layer 2).

### 3.2.3   STAGE 3: SOFTWARE DEVELOPMENT

The next stage of this research is translating the ISF that has been incorporated with ISO 27001 components into application software (ISM) to help users review their information security status and circumstances and compare it with the ideal situation that has to be satisfied by organizations (Figure 3.1, layer 3).

The software development methodology adopted to reconcile, translate and develop the user requirements based on ISF algorithms is visual object oriented programming (VOOP/Visual Basic) language for development of the graphical user interface (GUI), visual database (Access) for data storage and structure query language (SQL) as a bridge between GUI and database for further update and data retrieval. In the software development stage, we refer to the water fall (WFA) and spiral model (SM) software development methodologies. At every development step, requirements, analysis, design-implementation and testing-evaluation were conducted to ensure every stage was "bug-free". For the releases of subsequent software versions, we used a "release management" approach where the software is released in stages as feedback is collected from the users. The latest ISM version is the final release to be installed and tested in the respondent organization's site.

ISM will be used to conduct RISC investigation and analyze security events in real time and to collect, store, and report for compliance to ISO 27001. ISM is consists of two main functions of ISM: security assessment management (SAM/e-Assessment) and security monitoring management (SMM/e-Monitoring). Each of these functions and the architecture of ISM will be explained further in Chapter 5.

### 3.2.4   STAGE 4: SOFTWARE DEPLOYMENT AND TESTING

The ISM deployment (Figure 3.1, layer 4) includes RISC investigation for ISO 27001 compliance, and testing of the SP/SQ. This stage of testing would take approximately 2-12 months. There are three clusters to be dealt for testing: ISO 27001 holders (cluster-I), ISO 27001-ready (cluster-II), and ISO 27001 consultants (cluster-III).

In RISC investigation, the user seeks an electronic measurement to reflect the organization's circumstances regarding the standard (Figure 3.1, layer 5). There are a number of parameters used in RISC measurement: 6 domains, 7 clauses, 21 controls, 144 assessment issues and 256 refined questions; they are essential components in the information security standard ISO 27001. This RISC measurement uses a 5-point Likert scale to portray the degree of the ISO's controls implementation. The Likert scale points used in the measurement are: "0" = no implementation, "1" = below average, "2" = average, "3" = above average and "4" = excellent.

SP/SQ measurement phase is the last stage of research. At this stage, the user is required to assess ISM as guided by eight fundamental parameters (8FPs – explained in previous chapter) as indicators of performance and quality.

Moreover, to measure SP/SQ aspects, a 5-point Likert scale is also used to obtain the user's valuations and recommendations for its features and usability. The scale values ISM's usability in assisting organizations to comply with ISO 27001 as such: "0" = not recommended, "1" = partially recommended (below average), "2" = recommended (average), "3" = mostly recommended (above average), "4" = highly recommended (excellent).

## 3.3  RESPONDENTS

We used *systematic and selected sampling,* also called the Nth name selection technique as suggested by Patton (2005), instead of random sampling. It is considered superior to random sampling because it reduces sampling error. The selection of respondents was carried out by systematic and selected sampling criteria, which was chosen by specific consideration and purpose. PWC (Price Waterhouse Cooper consultants) conducted an information security breaches survey (ISBS) 2008, 2010, 2012, which indicated that different businesses have different criteria and attention given to issues related to information security and social engineering attacks. Based on the PWC findings, we selected ten organizations which can be classified into 6 clusters as our respondents to help us improve and refine the framework (ISF) and to test the application software derived from the framework (ISM). The respondents (organizations) were from the telecommunication, banking, finance, airlines, healthcare and ICT consultants. The majority of those companies are listed on the stock exchange, and the companies are well recognized by their clients as well as public. The further cluster of respondents will be discussed shortly in Subsection 3.3.3.

### 3.3.1  RESPONDENTS CRITERIA

These are the main reasons and criteria to why these 10 organizations were selected for this study:

1.  These organizations have serious attention on information security. Either on the information related to their business processes, customers or on their intellectual property rights with respect to products or services they provide.

2.  These organizations are willing to share information associated with their information security strategies, management, ISO 27001 compliance stage and accidents or attempts for security breaches.

3.  The organizations do not have any objection to the testing and installation of the system (ISM) in their business environments. This is a very important issue as ISM's installation might affect the performance of the main computer network. Moreover, ISM testing also consumes a lot of time since this testing includes the following processes:

    a.  To encompass all the controls in the standard – a total of 21 controls, 144 assessment issues and more than 256 refined questions were tested. Therefore, we required respondents who were able to follow and conduct all the processes of the research stages and allow RISC investigation and SP/SQ measurement.

    b.  To learn and understand how to assess the controls, assessment issues and refined questions conducted by the ISM.

    c.  To understand how the system works in real-time for monitoring the potential suspects and intruders in the computer network.

4.  Time constraints. With a 3-year PhD research, there is a need to be well-planned and well-scheduled with the proper respondents who are cooperative and pay considerable attention to the research. Therefore, we chose 10 organizations which include the six business clusters as mentioned earlier.

## 3.3.2  RESPONDENT ORGANIZATIONS' SIZE

PWC defined the sizes of the organizations in their survey of information security. Small organizations have a staff of less than 50 people, medium organizations have a staff between 50 and 250 people and large organizations have a staff of more than 250 people in a building or area of the organization (Potter & Beard, 2010, 2012).

TABLE 3.1   The Respondents' Business Fields and Size

| Respondent's main business area | Number of organization(s) | Number of employee |
|---|---|---|
| Automotive & Manufacturing (large organization) | 1 | > 250 |
| Banking Regulator (large organization) | 1 | >250 |
| Financial Service (large organization) | 1 | >250 |
| Telecommunications (large organization) | 1 | >250 |
| Airlines (large organization) | 1 | >250 |
| ICT Consultant (small organization) 50–250 (medium organization) | 2 | <50 |
| Health Centre (small organization) | 2 | <50 |
| Research Institute (small organization) | 1 | <50 |

We did the test to the three sizes of organizations following the same groupings as PWC, considering that the size of the organization affects its vulnerability to information security. Large organizations are more susceptible to being attacked and compromised by a hacker. The respondents' profiles regarding the sizes of their organizations are shown below in Table 3.2.

### 3.3.3  RESPONDENTS CLUSTER

The selected organizations were grouped into three clusters according to their stage of compliances with ISO 27001 (Table 3.3). We intend to

**TABLE 3.2**  Respondent Organizations' Profiles

| Companies Profile | | | | |
|---|---|---|---|---|
| In what sector was each respondent's main business activity? | What standards have respondents complied with? | What percentage of IT budget was spent on IS | Do you need a consultant, in helping you to understand the IS standard? | Respondent's Market Share within their segmentation |
| Automotive & Manufacturing | CISCO & Microsoft | 11–25% | Yes | 30–35% (2) |
| Banking Regulator | ISO 27001 (since 2007) | >25% | Yes | Regulator |
| Telecommunications | ISO 27001 (since 2012) | >25% | Yes | ~50–60% (1) |
| Airlines | ISO 9001, ISO 14000 (ongoing process ISO 27001) | 11–25% | Yes | ~45–55% (1st-national) (2nd ASEAN after SQ) |
| Research Institute | COBIT & ITIL (ongoing process ISO 27001) | 11–25% | Yes | 15% |
| Financial Service | Microsoft Certificate (ongoing process ISO 27001) | 11–25% | Yes | 25–30% |
| ICT Consultant | ISO, ITIL, Microsoft | >25% | Not | ~5–10% |

**TABLE 3.3**   Person in Charge At the Respondent Organizations

| Respondent | Person in charge | Level |
|---|---|---|
| Automotive & Manufacturing | • IT manager<br>• IT officer | • Medium level of technical expertise<br>• Medium level of management |
| Banking Regulator | • Information security officer<br>• ISO 27001 officer | • Senior level of technical expertise |
| Financial Service | • Risk manager | • Medium level of management |
| Telecommunications | • Business portfolio and synergy data support officer<br>• ISO 27001 implementation kick off and live test evaluation team | • Medium level of management<br>• Medium level of technical expertise |
| Airlines | • Scheduling, operational and maintenance manager | • Senior level of management |
| ICT Consultant | • System analysis<br>• Network and Security engineer | • Senior level of management<br>• Medium level of technical expertise |
| Health Centre | • Medical doctor | • Medium level of management<br>• Medium level of technical expertise |
| Research Institute | • Programmer/Developer | • Senior level of technical expertise |

determine patterns and strategies for obtaining certification from cluster-I. Organizations in cluster-II were given ISF-ISM to be used as a tool for measuring the RISC and organizations in cluster-III were expected to provide views and advice on whether ISF-ISM will significantly help organizations to comply with ISO 27001. Those clusters are expanded as given in the following subsections.

### 3.3.3.1   Cluster-I: ISO 27001 Holders

This cluster consists of companies that recently received certification by ISO 27001 in the period of 2010–2012, or their certification is still valid at the time this study was conducted.

### 3.3.3.2   Cluster-II: ISO 27001 Ready

This cluster consists of companies who are currently pursuing ISO 27001 compliance, whether they are in document preparation stage, scenario development stage, or risk management analysis stage.

### 3.3.3.3   Cluster-III: ISO 27001 Consultants

This cluster consists of ICT consultants in the security area, particularly information security, assessment and standards.

### 3.3.4   *RESPONDENTS' KEY-PERSON(S)*

From the 10 selected organizations, we chose key persons from each organization as representatives (they are potential users) that would be competent in providing input, feedback, and advice on behalf of the organization. These respondents were the employees who were directly in charge of information security and related tasks. They have very good knowledge and skills on information security. These respondents have huge responsibilities to protect the information systems assets of their organizations. The following are the titles and roles of these respondent key person(s) (Table 3.4):

1. *Information security officer and ISO 27001 officer* (infosec officer). We discussed with the infosec officer from the regulator bank (Central Bank as Authority and Monetary Agents), to draw from their experiences about the organization's journey to obtain an ISO 27001 certification. The main tasks of the infosec officer are to follow the rules of the infosec tripartite, CIA, and record infosec breaches that occur within the organization. In addition, as the regulator, the organization also provides guidance and regulation to other financial institutions on how to comply with ISO 27001 to ensure information security within the organizations concerned.

**TABLE 3.4**   Nominal Style Scalet

| Gender | : ☐ Male | ☐ Female | | |
|---|---|---|---|---|
| Age | : ☐ < 25 | ☐ 25–35 | ☐ 35–45 | ☐ >35 |
| Education Level | : ☐ Diploma | ☐ Bachelor | ☐ Master | ☐ PhD |

2. *Security systems analyst and developer* (SSAD). SSADs function as consultants for IT information security solutions by providing advice on how their clients could satisfy ISO 27001's requirements. An SSAD needs to ensure that every application, common software and custom-made software are well developed and meet the requirements of the chosen information security standard.

3. *IT manager and IT Officer* (ITMO). The main task is to handle or oversee the company's technology, computer networks, Internet and local networks. An IT manager also directly supervises the system at the regional level. In his capacity as a manager, he typically oversees the information technology hardware and software to ensure a secure network.

4. *Business portfolio and data synergy support.* ISO 27001 implementation kick off and live test evaluation team. The team involved in preparation for certification that deals with information security scenarios, such as infrastructure management programs, disaster recovery systems, operations, and maintenance.

5. *Risk manager.* This person is responsible for managing risk for the respondent organization (financial institution/bank). An organization needs to ensure information resources, electronic services and transactions are well protected to an acceptable level of information security risk. Risk management (RM) is part of ISO 27001. It is mentioned in the clause of business continuity management and exists in the control of information security in the business continuity process and business continuity risk management/ assessment.

By identifying and proactively addressing risks and opportunities, an organization protects and create value for their stakeholders, including owners, employees, customers, regulators, and the society overall. RM has gained considerable importance among organizations because of the increased interdependence between businesses, regulatory demands and a growing awareness of its importance in preventing systemic failure of information security. In other words, companies must demonstrate that they actually use RM as a key component of business processes, particularly associated with information security.

6. *Scheduling, operational and maintenance managers.* In the airlines industry, securing information is a high priority. Aircraft maintenance, the accuracy of flight schedules, take-off and landing times, passenger manifests and synchronisation with the cabin-crews on duty. It is a combination of operations needing information security that should be maintained at the highest level as possible with zero tolerance.

7. *Medical doctor* (MD). Responsible for maintaining their information especially the security of patients' information such as patient data, drugs, or other matters relating to patient habits.

## 3.4   DATA COLLECTION METHOD

We conducted data collection from the respondents as described in the previous section. Data collected primarily related to expectations, tendencies, assessments, and perspectives of research objectives and its stages. From these data we can gather the respondents' feedbacks on the proposed framework, features and specifications of the software, circumstances of the respondent organization's information security and finally the respondents' perception of the performance and quality of the software.

The time taken to complete all stages of data collection: brainstorming, filling questionnaires, user requirements, testing-running and evaluation of the system was 18 months. The most time consuming part was performing the comprehensive ISF-ISM RISC testing, since there were many aspects that were assessed; 144 assessment issues, 21 controls, 7 Clauses, and 6 domains of ISO 27001.

The following data collection methods are employed: discussion, survey, testing, and assessment. The details of these methods are discussed in the following subsections.

### 3.4.1   FOCUS GROUP DISCUSSION (FGD) AND OPEN QUESTIONS

Brainstorming and sharing sessions conducted with respondents. The main objective of FGD here is to gather people from similar backgrounds or experiences together to discuss information security topics of interest. Each session is guided by a moderator (or group facilitator) to introduce information security topics for discussion. The advantages of FGD are allowing the respondents express their opinions which provide an insight to how respondents think about information security issues and compliance.

In this research, we used FGD sessions to collect information from practitioners about the experiences, security breaches, document preparation for information security standard implementation and success stories on compliance with ISO 27001 certificates.

Discussions were guided by open-ended questions, conducted face-to-face with the respondents in order to obtain feedback, opinions and advice in relation to the information security issues, business processes, information security strategies, information security for company branding and information security breaches that have occurred in the organization, and also how organizations prepare things to meet the requirements of ISO, such as:

1. Document preparation.
2. Scenario in the face of information security threats.
3. Security infrastructure.
4. Challenges and difficulties in meeting the existing controls.
5. Barriers in understanding the technical terms in the standard.
6. The length of time required.
7. Consultants and an expertise to support organizations.
8. Human resources that must be prepared to support compliance stages.
9. Support from the management level (director and commissioner).
10. The need for an approach or framework to facilitate and shorten the time required to fulfill existing controls.
11. Respondent's necessity for self-assessment and RISC evaluation to comply with ISO 27001 certification through a software application.

### 3.4.2 QUESTIONNAIRE

Questionnaires were distributed to respondents to collect information on security related issues, functions and technical requirements from users (respondents). Items in the questionnaire were developed on the basis of infosec issues. The questionnaire was divided into two parts. The first part was distributed in the first stage of research. The main objective was to get feedback from potential users on the need of a framework and tool for ISO 27001 compliance. The second part contains measurements for ISF and ISM. The objective of the second part was to perform RISC investigations for ISO 27001 compliance and appraisal of ISM's SP/SQ.

### 3.4.2.1   Questionnaire Structure

The questionnaire was divided into four main sections: attitudes to information security, security awareness, security standards and framework/comprehensive evaluation tools. Each section consists of specific and essential questions to obtain user requirements and feedbacks on the proposed framework and tools. The details of the structure of the questionnaire are as follows:

- *Attitudes to information security.* This section consists of an overview of the respondent organizations' attitudes and perceptions on their information and how to maintain its security. It consists of six questions that are associated with:
    1. Respondents' main business activities.
    2. Respondents' role within their organization.
    3. Respondents' businesses which are currently carrying out e-business activity.
    4. The web site activity and function within respondent organizations.
    5. Importance of information security assets to support and enable respondents' business processes.
    6. Priority level of information security.
- *Security awareness.* All questions within this section represent the knowledge and attitudes that members of an organization possess with regards to the protection of the physical entities, especially those in relation to information assets of the respondent organizations. Many organizations require formal security awareness training for all workers when they join the organization, which is

conducted periodically, usually annually. It consists of 11 questions associated with:

1. Main driver for information security expenditure.
2. Perception on security incidents next year.
3. Perception on the difficulties of catching security breaches in the future.
4. Level of the worst security incident suffered.
5. Importance of reporting incidents.
6. How to prevent future incidents.
7. Documented the security requirements.
8. Percentage of IT budget for information security.

- *Security standard.* This section covers regulation and standard issues that govern information security within an organization. Information security has become increasingly important in the information era and recognized as a key asset by many organizations. It has become a business enabler and an integral part of a business to raise the trust of customers and to effectively use emerging technologies for the business process. Organizations are paying increasingly more attention to information protection by implementing an ISMS standard. The governing principle behind the standard is that organizations should design and be aware of such information security scenarios; implement and maintain a coherent set of policies, scenarios and implementations to manage risks, vulnerabilities, and threats to its information. This section also obtains user expectations and feedback related to the new methodology, framework and tool, as an approach for satisfying the standard's requirements. It consists of 12 questions associated with:

1. Advantages of implementing an information security standard.
2. Type of information security standard to be used on respondent organizations.
3. How long it takes to implement information security standards.
4. Obstacles and challenges faced during ISO 27001 implementation.
5. How difficult it was to understand the terms, concepts and controls of ISO 27001.
6. How long it takes to understand the terms, concepts and assessment issues.

7.  The need for third party consultants or expertise in helping respondent organizations to understand the terms, concepts, and controls.

8.  How respondents think the new framework (ISF) can assist them in understanding the terms and concepts of information security standards.

9.  Benefits of introducing ISF.

10. Respondent's expectation on ISM for RISC investigation.

11. Features expected for ISM.

- *Framework and Software evaluation.* This section aims to measure the quality and performance of the framework and software (SQ/SP), which refers to how effective ISM functions to help users in understanding and measuring the organization's RISC. Here we used the five point Likert scale to evaluate the SQ/SP of ISM:

  - *0: not recommended (not implemented);*
  - *1: partially recommended (below average);*
  - *2: recommended (average);*
  - *3: highly recommended (above average);*
  - *4: excellent.*

This section consists of 12 questions that are associated with:

1.  Perception on how ISF helps respondents understand the information security standard terms and concepts.

2.  The benefits that respondents got after using ISM.

3.  Time taken to assess the readiness level for organization's information security standards implementation using ISM.

4.  Domain concerned.

5.  ISM features that aid an organization's understanding of its security conditions.

6.  The importance of implementing an information security standard in the organization.

## 3.4.2.2   Questionnaire Design

The questions in the questionnaire were deliberately designed to be friendly and to garner better response rates by only requiring the respondents to tick their preferred choice of answer. The descriptive nature and quantitative approach of this study dictated the design of the questionnaire.

The questionnaire mostly comprised of a series of statements followed the five-point Likert rating scale. The questionnaire was designed primarily to answer the research questions and objectives mentioned earlier.

The demographic section of the questionnaire focused on generating data using a nominal scale from the respondents. It gives some basic, categorized and gross information of the respondents to calculate the frequency or the percentage of each category. An example is as shown in Table 3.4.

An ordinal scale was extensively used in this research in order to determine the percentage of respondents who consider the items listed to be important. This also enabled ranking orders to be obtained. An example is as shown in Table 3.5.

An interval scale was also extensively used in the questionnaire. This scale is powerful, since it can measure the magnitude of the differences in the preferences among the respondents as well as the central tendency, dispersion, standard deviation and the variance when such information are needed.

The 5-point Likert scale is extremely popular for measuring attitudes, perceptions, feelings, emotions and attitudes of respondent organizations. In this case, the measurement of the attitudes, perceptions, feelings and emotions of the respondents would be obtained through the respondents' responses in accordance to their degree of agreement and disagreement with the predetermined statements. An example is shown in Table 3.6.

**TABLE 3.5** Ordinal Scale Style

What is the *main driver for information security expenditure?*

(Please tick [] the appropriate boxes. You may tick more than one box)

☐ Protecting customer information;

☐ Preventing downtime and outages;

☐ Complying with laws/regulations;

☐ Protecting the organization's reputation;

☐ Maintaining data integrity;

☐ Business continuity in a disaster situation;

☐ Protecting intellectual property;

☐ Enabling business opportunities;

☐ Improving efficiency/cost reduction;

☐ Protecting other assets (e.g., cash) from theft.

TABLE 3.6  The 5-Point Likert Scale Style

| S. No. | Statements | 4 Excellent | 3 Highly recommended (above average) | 2 Recommended (average) | 1 Partially recommended (below average) | 0 Not recommended (not implemented) |
|---|---|---|---|---|---|---|
| 1 | At what level you are recommending i-solution modeling software for information security self-assessment (RISC)? | | | | | |
| 2 | At what level you are measure the benefit of this i-solution modeling software to help you understanding ISO27001 controls | | | | | |
| 3 | At what level you are measuring i-solution modeling software in order to understanding information security standard term and concept? | | | | | |
| 4 | At what level you are measuring features in this software | | | | | |
| 5 | At what level you are measuring user interface and easiness level of this software | | | | | |
| 6 | At what level you are define of analysis precision produced by this software | | | | | |
| 7 | At what level you are state on final result precision produced by this software | | | | | |
| 8 | At what level you are indicate on performance provided by this software | | | | | |

## 3.4.3 ISM RISC INVESTIGATION AND SQ/SP MEASUREMENT

To test ISM for a comprehensive evaluation, RISC and SP/SQ, ISM has to be installed in the respondent organization's site to measure the level of readiness and information security capabilities (RISC) and monitor potential suspects of security breaches. Once installed, ISM can be used by the user (defined as the special respondent that would be using an ISM as a measurement tool) to investigate the RISC level, as required by ISO. Each respondent was also asked to assess ISM performance based on defined 8FPs (eight fundamental parameters). Those parameters are the features which they expected as obtained from feedback in the user requirements stage.

Users need to learn the terms and concepts used by ISO 27001. This can be done easily through features provided by the software as these features are designed to assist users to understand the concepts and technical terms of ISO 27001 easily. Once users are familiar with the concepts and terms, they could perform measurements of their organization' RISC through filling the electronic evaluation form provided by ISM. This electronic evaluation form consists of the assessment issues, controls, clauses, and domains as dictated by ISO 27001.

**TABLE 3.7** Assessing the ISO 27001 Control Concerned with "Organisation – Organisation of Information Security – Allocation of Information Security Responsibilities"

| Assessment Issue | Refined Question | Answer |
|---|---|---|
| Objects of responsibility | Are the assets and security processes clearly identified? | 2 |
| Levels of responsibility | Does the position of information security manager that takes the overall responsibility exist? | 3 |
| | Are the authorization levels identified? | 2 |
| | Does each asset have a responsible person? | 2 |
| | Does each security process have a responsible person? | 3 |
| Documenta-tion | Does reporting of the above exist? | 1 |
| Achievement Result | | 2 |

As an example, for the RISC measurement step, assessment parameters are described as follows (Table 3.7): ISO 27001 domain: organization; Control: Organization of information security – allocation of information security responsibilities; *Assessment* issue: Are assets and security process clearly identified?

The stakeholders should interpret ongoing situations, implementation and scenarios in the organization, and benchmark it to the security standard level of assessment as a reference standard. ISM gives an overview of the results level-by-level, starting from the lowest level (control), middle level (domains) and the highest level (top domains). It provides feedback, a score result, and a strengths-weaknesses analysis associated with the steps and scenarios to be conducted by the respondent organization. The strengths-weaknesses analysis feature provides a convenience review to the users and management to catch up and reveal the weaknesses of the organization in an efficient and focused method. It is due the user only focusing on the points that need to be improved.

## 3.5  DATA CATEGORIES

As is clearly mentioned in the section research stage, our study consists of four phases of data collection as input and consideration for the development of the framework. Consequently, there are four categories of data and they are discussed in the following subsections.

### 3.5.1  *FOCUS GROUP DISCUSSION (FGD) DATA*

The outcome obtained from the FGD stage in sharing the organization's experiences on information security. These are about the experiences of ISO 27001 certificate holders such as how organizations prepare to meet the requirements of ISO:

  i.   documentation;
 ii.   scenario;
iii.   infrastructure;
 iv.   challenges, barriers and difficulties;

v.   time consumed;
vi.  human resources issues related to information security;
vii. management support;
viii. respondent expectations of a new framework;
ix.  respondent needs for a software application.


## 3.5.2  QUESTIONNAIRE DATA

Data from this questionnaire was the key indicator of the respondent organizations' trends, plans, expectations and requirements, which is a continuation of the previous stage by FDG. All those results were used as a guideline to propose and develop the ISF and ISM. Requirements, features and technologies expected by the users are as shown below:

i.   Importance of proposed framework to understand the terms and concepts of ISO 27001;
ii.  Formula to measure readiness level;
iii. Strengths-weaknesses analysis to review information security circumstances;
iv.  Software as an application of the framework to measure and assess readiness level of the organization for the implementation of an information security standard;
v.   The benefits possibly gained from the framework and software.


## 3.5.3  RISC INVESTIGATION DATA

The data generated in this phase is an example of RISC measurement by respondents. They assign values to all parameter controls provided, and ISM organises measurement results in each level starting from controls, clauses, domains and top domains. All those measurements are featured by a strengths-weaknesses analysis as a reference for the organisations' improvement. This measurement follows the ISF formula (more details will be discussed in Chapter 4: Proposed Framework), which is expressed as follows:

$$x_h = \sum_{1}^{6} \frac{\left[\sum_{i=1}^{n} \frac{\left[\sum_{j=1}^{n} \frac{\left[\sum_{k=1}^{n} \frac{[assessment\ issues]_k}{n}\right]_j}{n}\right]_i}{n}\right]}{6}$$

where $x_h$: top domain; $i$: domains; $j$: clauses; $k$: controls.

### 3.5.4   ISM SQ/SP MEASUREMENT DATA

ISF SQ/SP measurement functions as a key performance and quality indicator of the ISM in executing RISC testing. The data is generated in this phase to demonstrate the level of quality and performance, and user acceptance of the technology and features offered by the ISM. The parameters of SP/SQ follow the 8FPs through McCall taxonomy (more details will be discussed in Section 3.6). The taxonomy could be express as follows:

$$F_a = \sum_{i=1}^{n} \frac{w_i c_i}{n}$$

where $F_a$ is the total value of factor $a$; $w_i$ is the weight for criterion $i$; $c_i$ is the value for criterion $i$.

### 3.6   DATA ANALYSIS

The data collected from the questionnaires were edited, coded and encoded to be analyzed using macro spreadsheet software. After the data had been collected through the questionnaires, they were immediately checked for consistency and completeness. For example, item non-responses like neglecting to answer some items or wrongly answered were rectified and edited. In accordance to Patton (1980), item non response of more than 25% of the questionnaire were discarded.

On the other hand, the results of testing and evaluation of ISF-ISM were further analyzed using software quality and performance parameters (SQ/SP-P) and release and evaluation management (REM) to find out ISF-ISM performance and quality as a tool to measure an organization's readiness level and information security capabilities (RISC). The quality of the software (software quality, SQ) and performance of the software (software performance, SP) is the theme of study and research in the history of the science of heredity software engineering (Wahono, 2006). It began with the object to be measured, processor product, how to measure the software, the measuring point of view and how to determine the parameters of software quality measurement. The first question that arises when discussing the measurement of SQ/SP is exactly which aspects are to be measured. SQ/SP can be viewed from the perspective of the software development process and the results of the resulting product. The final assessment is oriented to how the software can be developed as expected by the user. This departs from the notion of quality according to IEEE's standard glossary of software engineering technology (IEEE, 1990) that states: *"The degree to which a system, component, process meets customer or user needs or expectations"*.

From the product point of view, McCall proposed taxonomy as the best practice of software measurement (McCall, 1977a, 1977b). The series of parameters to measure software performance are: software advantages, accuracy, precision, stability, ease of use, user interfaces. In this study, the author adopted, refined, and improved SQ/SP parameters to become the eight fundamental parameters (8FPs) as a function to measure performance the software ISM. All 8FPs reflect ISM performance, behavior, function, and back-office programming to handle the workload of RISC investigation and monitoring. The details and descriptions of the parameters as ISF-ISM performance indicators, together with the defined scale associated with 8FPs are shown in Table 3.8.

Moreover, SQ/SP can be measured quantitatively to assess 8FPs. The measurement processes of 8FPs following the McCall taxonomy (McCall, 1977a, 1977b). This taxonomy follows a hierarchic structure, where the upper-level (high-level attributes) are called factors and the lower level (low-level attributes) are called criteria. The McCall taxonomy formula is explained as follows:

**TABLE 3.8** Performance Parameters (8 Fundamental Parameters – 8FPs)

| ISM SP/SQ indicators | Description |
|---|---|
| 1. The benefit of ISM to help organization understanding ISMS standard's | Through ISM, the organization will be assisted in understanding the details of the standard, so that the need for consultants can be minimized. |
| 2. Final result precision produced by ISM | Feature that describes the final result of a RISC investigations. It is consists of 6 domains and 21 essential controls. This result is shown in detail together with the strong-point histogram. |
| 3. Analysis precision produced by ISM | To generate proper analysis using strength-weakness points. The organization could focus on weak points for further improvement. |
| 4. ISM Performance | Performance of ISM as a whole, the handling and completing of mathematical calculations, as well as real-time monitoring. |
| 5. ISM features | Assessing the features possessed by the ISM in conducting RISC investigations and real-time monitoring. |
| 6. ISM graphical user interface and user friendliness | ISM ease of operation. It is supported by the interface, design, and layout of the menu that is accessible and understandable. |
| 7. ISM can be used to understand information security standard terms and concepts | ISO 27001 consists of technical terms and concepts. Hence the need for a tool to easily understand those terms and help make decisions for security standard adoption. |
| 8. ISM functions as information security self-assessment | Assessing the functions of self-assessment as organization enhancement to the information security standard. |

$$F_a = \sum_{i=1}^{n} \frac{w_i c_i}{n} \tag{1}$$

$$F_a = \frac{w_1 c_1 + w_2 c_2 + w_3 c_3 + \ldots + w_n c_n}{n}$$

where $F_a$ is the total value of factor $a$; $w_i$ is the weight for criterion $i$; $c_i$ is the value for criterion $i$.

Stages that must be followed in the measurement are:

- Stage 1: Define the criteria to measure a factor;

- Stage 2: Determine the weight (w) of each criterion (e.g., →1; if all criterion are in same level affected to the whole performance of the software);
- Stage 3: Determine the scale of the value criteria (e.g., →$0 <= c <= 4$);
- Stage 4: Give the measurement value of each criterion;
- Stage 5: Calculate the total value of the formula [1].

## KEYWORDS

- **Likert scale**
- **McCall taxonomy**
- **readiness and information security capabilities**
- **risk management**

**CHAPTER 4**

# INTEGRATED SOLUTION FRAMEWORK

## CONTENTS

## 4.1  INTRODUCTION

The business environment is changing rapidly, and organizations are becoming increasingly interconnected and transmit their information through the Internet. The changing business environment is creating new vulnerabilities. As a consequence, an information security management system is the answer to support and enable activities of the organizations. On the other hand, organizations frequently encounter difficulties in complying with information security standards. According to Standish Group (2013), many information security standardizing and ISO 27001 compliance projects were failures and caused losses of billions of dollars,

and only around 13% of ISO 27001 standardizing projects were successful. There were several major problems that caused the delay for ISMS
and ISO 27001 projects, such as technical barriers, the project owner's
absence of understanding, technical aspects, lack of internal ownership,
and neglect of certain aspects.

One of the key components for the success of information security
certification is by using a framework. This framework acts as a tool to
understand the processes and technical aspects. Unfortunately, existing frameworks do not provide formal and practical models for RISC
measurements. The proposed framework is developed to overcome the
main shortcomings of existing frameworks including providing this
formal and practical model for RISC measurement. This framework is
also designed in such a way so that we can derive an integrated solution to overcome the organization's technical barriers and difficulties
in understanding, investigating and complying with the ISMS standard
(ISO 27001).

This chapter will focus on one of the key steps taken towards successful adoption of the new framework. It begins with an analysis, opportunities, presentation, and discussion of the existing frameworks to find out
respective strengths and weaknesses as our foundation for proposing a
new framework. Then we will discuss the ISF structure, mathematical for-
mulations, computer algorithms, strengths-weaknesses analysis and effectiveness of ISF for RISC investigation.

## 4.2  THE CHALLENGES

The main challenge tackled by this study is the fact that existing frameworks do not provide formal and practical models for RISC investigation, therefore it is our intent to propose a framework that could integrate
formal and practical models for RISC investigation. This framework is
designed in such a way to become an integrated solution to overcome the
organization's technical barriers and difficulties in understanding, investigating and complying with ISO 27001.

Compliance to a standard is the process of comparing the applied controls of an organization with those in the standard. The compliance process

is also called *information security assessment*. It is basically a gap analysis in which the differences between an organization's information security circumstances with the standard are discovered. Checking conformity level helps an organization to determine its relative position to the standard, which is useful for the certification process.

According to Kosutic (2010, 2013) the information security assessment is probably the most complex and challenging part of ISO 27001 implementation. However, it is the most important step towards adoption of ISO 27001, as this assessment can find any potential security gaps (i.e., assess the risks) (Alfantookh, 2009) and address the most appropriate ways to avoid such incidents by referring to the associated controls (i.e., treat the risks) (Calder & Watkins, 2010, 2012).

There are six basic steps to assess risks in relation to ISO 27001 (ISO, 2005; Kosutic, 2013), namely (1) risk assessment methodology; (2) risk assessment implementation; (3) treatment implementation; (4) information security assessment report; (5) statement of applicability; and (6) risk treatment plan. The details of each step are explained in the following subsections.

## 4.2.1 RISK ASSESSMENT METHODOLOGY

An organization needs to define rules on how to perform risk management, since the biggest problem with assessment happens if different parts of the organization perform assessment in a different way. Therefore, an organization needs to define whether they want qualitative or quantitative risk assessment, which scales will be used for qualitative assessment and what will be the acceptable level of risk.

## 4.2.2 RISK ASSESSMENT IMPLEMENTATION

Once the rule is fixed, an organization can start finding out what potential problems could happen, list all assets, then threats and vulnerabilities related to those assets, assess the impact and likelihood for each combination of assets, threats, vulnerabilities, and finally calculate the level of risk.

### 4.2.3   TREATMENT IMPLEMENTATION

This stage is where an organization needs implements the solution considering, for instance, how to decrease the risks with minimum investment by transferring the risk to another party, e.g., to an insurance company by buying an insurance policy. Risks can be minimized by stopping an activity that is too risky, or by doing it in a completely different fashion. Accept the risk, for instance, if the cost for mitigating that risk would be higher than the damage itself.

### 4.2.4   INFORMATION SECURITY ASSESSMENT REPORT

**An information security assessment report** is documentation of all scenarios and controls that have been done for auditor checks, also for an organization to review for their improvement.

### 4.2.5   STATEMENT OF APPLICABILITY (SOA)

This stage shows the security profile of an organization, based on the results of the risk treatment and risk management, where the organization lists all the controls it has implemented. The SoA is the central document that defines how an organization will be (or has been) implementing information security. It is the main link between the risk assessment and treatment and the implementation of information security, whose purpose is to define which of the suggested 133 controls, including 21 essential controls of security measures, will apply, and for those that are applicable, how they will be implemented. Ordinarily, if an organization goes for ISO certification, the auditor takes the SoA report and checks whether the organization has implemented the controls in the way described in the SoA.

### 4.2.6   RISK TREATMENT PLAN

The Risk Treatment Plan defines who is going to implement each control and in what specific scenario, in which timeframe, with which budget.

Risk treatment plan also called *implementation plan* or *action plan* (Kosutic, 2013). Once the ISO 27001 certification project manager prepares the document that covers all those stages, it is crucial to get management approval because it will take considerable time, effort, and money to implement all the controls planned. Then the respondents start the journey from not knowing how to set up effective information security measures, to having a very clear picture of what needs to be implemented. ISF will help to make this journey systematic.

ISF applies to all respondent clusters – ISO 27001 holders and ISO 27001 ready – where the role of organizations are to passively participate in the certification process by following what consultants evaluate during certification stages and also actively as partner by enhancing their ISMS ability and literacy. On the other hand, for ISO 27001 consultants, the proposed framework could assist them to provide their clients with a guidance track to accelerate the certification process.

## 4.3   THE MOTIVATIONS

In this study, there are several substantive reasons that motivate the development of ISF. Firstly, ISF is the result of an extensive literature review also justified by respondent feedback that state that a framework for formal RISC measurement is needed. To develop the framework, this study is built upon recent reviews of frameworks and tools that have been developed for this purpose. The literature survey observed points of concern, advantages, and disadvantages of the existing frameworks (called by the nine state-of-the-art frameworks or 9STAF). The 9STAF are widely used in relation to information security issues. The 9STAF were studied carefully and a comparison among them was carried out to reveal their salient features and respective positions.

Secondly, as an academic contribution to the scientific and practical world by proposing a framework, then for future research, the framework could then accommodate and customize ISF to fit with other standards such as BS 7799, COBIT, ITIL, and others. ISF could be implemented to these other standards by mapping the stages through grouping of controls to respective domains in each standard.

Finally, to introduce a novel algorithm for compliance measurement and investigation of ISMS as a bottom–up approach, designed to be implemented in high-level computer programming language, to produce a graphical user interface (GUI) that is easy to use and powerful for ISO 27001 investigation.

## 4.4 EXISTING FRAMEWORKS: THE NINE STATE-OF-THE-ART FRAMEWORKS (9STAF)

There are nine state-of-the-art frameworks (9STAF) available to help, abstract and organize efforts to comply with information security standards. Those frameworks are: (1) a Framework for the Governance of Information Security (Posthumus and Solms, 2004); (2) a Framework for Information Security Management based on Guiding Standards: a United States perspective (Sipior and Ward, 2008); (3) a Security Framework for Information Systems Outsourcing (Fink, 1994); (4) Information Security Management: a Hierarchical Framework for Various Approaches (Eloff and Solms, 2000); (5) Information Security Governance Framework (Ohki et al., 2009); (6) Queensland Government Information Security Policy Framework (QGISPF, 2009); (7) STOPE methodology (Bakry, 2004); (8) a Security Audit Framework for Security Management in the Enterprise (Onwubiko, 2009); (9) Multimedia Information Security Architecture Framework (Susanto & Muhaya, 2010). Table 4.1 shows briefly the profile and features of the frameworks.

The Framework for the governance of information security (FGIS) which was introduced by Posthumus and Solms (2004) reveals and suggests the important part of protecting an organization's vital business information assets. This action should be considered as such and should be included as a part of corporate governance responsibility by the corporate executives.

Sipior and Ward (2008) introduced the second framework that is intended to promote a cohesive approach, which considers a process view of information within the context of the entire organizational operational environment.

Fink (1994) suggested a security framework for information systems outsourcing, a framework which assumes that most of information systems

**TABLE 4.1**   Existing Frameworks for Information Security

| Framework | Features and functions in brief |
| --- | --- |
| A framework for the governance of information security (Posthumus and Solms, 2004) | It states the importance of protecting an organization's vital business information assets by investigating several fundamental considerations of corporate executives' concern. |
| A framework for information security management based on guiding standard: a United States perspective (Sipior and Ward, 2008) | It is intended to promote a cohesive approach that considers a process view of information within the context of the entire organizational operational environment. |
| | The four levels of information security: international oversight of infosec[1], national oversight of infosec, organizational oversight of infosec, employee oversight of infosec. |
| A security framework for information systems outsourcing (Fink, 1994) | It assumes that most infosec related matter can be outsourced. Consequently, information security issues are controlled by the outsourced company. |
| | The aim of this framework is to evaluate the loss in infosec security and control when infosec outsourcing occurs. |
| Information security management: a hierarchical framework for various approaches (Eloff and Solms, 2000) | It elucidates ill-defined terms and concepts of infosec. |
| | All organizations are dependent on their infosec resources in today's highly competitive global markets, not only for their survival but also for their growth and expansion |
| Information security governance framework (Ohki et al., 2009) | This framework supports corporate executives to direct, monitor, and evaluate ISMS related activities in a unified manner |
| Queensland government information security policy framework (QGISPF, 2009) | This framework identifies various areas which contribute to effective information management and serves as an organizing framework for ensuring appropriate policy coverage and avoiding overlaps which may occur without such a framework |
| STOPE methodology (Bakry, 2004) | STOPE introduces domains of attention in information security to better focus the analysis of existing problems from the perspectives of Strategy, Technology, Organization, People and Environment |

**TABLE 4.1**   (Continued)

| Framework | Features and functions in brief |
|---|---|
| A security audit framework for security management in the cntcrprise (Onwubiko, 2009) | This framework comprises five components and three subcomponents. The components are: Security policy that defines acceptable use, technical controls, management standards and practices. |
| | Regulatory compliance stipulates acceptable regulatory and security compliances |
| Multimedia information security architecture framework (Susanto and Muhaya, 2010) | This framework emphasizes on information security in multimedia objects. Proposed a Multimedia Information Security Architecture which the authors based on ISA architecture with 5 main components, namely: *Security Policy, Security Culture, Monitoring Compliance, Security Program and Security Infrastructure.* |
| | MISA consist of 8 main components, namely: *Security Governance, Security & Privacy, Multimedia Information Sharing, Protection & Access, Security Assessment, Cyber Security, Enterprise Security, and Security Awareness* |

related matters can be outsourced. Consequently, information security issues are controlled by the outsourced company. The aim of this framework is to evaluate the loss in IS security and control for IS outsourcing.

The hierarchical framework for information security management introduces the principal aim to assist management in the interpretation, as well as in the application of internationally accepted approaches to IS management (Eloff & Solms, 2000).

Ohki et al. (2009) introduced the information security governance (ISG) framework, which combines and inter-relates existing information security schemes; Corporate Executives can direct, monitor, and evaluate ISMS related activities in a unified manner, in which all critical elements are explicitly defined along with the functions of each element, and interfaces among elements.

Queensland government information security policy framework (QGISPF) defines the generic classification scheme for information security policies, and does so with a perspective that it is independent from the physical implementation models chosen by departments, agencies, and offices. GISPF identifies and defines various areas, which contribute to effective information management and serve as an organizing framework for ensuring appropriate policy coverage and avoiding overlaps that may occur (QGISPF, 2009).

A security audit framework for security management in the enterprise comprises five components and three subcomponents. The components comprise of a security policy that defines acceptable use, technical controls, management standards and practices. It focuses not only on technical controls around information security, but also processes, procedures, practices and regulatory compliance that assist organizations in maintaining and sustaining consistently high quality information security assurance (Onwubiko, 2009).

Susanto and Muhaya (2010) introduced multimedia information security architecture framework which emphasizes on information security in multimedia objects and issues in existing architecture.

STOPE methodology introduces domains of attention in information security to better focus the analysis of existing problems from the perspectives of strategy, technology, organization, people and environment, explained further as such: *Strategy:* the strategy of the country with regards to the future development of the industry or the service concerned. *Technology:* the technology upon which the industry or the service concerned is based. *Organization:* the organizations associated with or related to, the industry or the service concerned. *People:* the people concerned with the target industry or service. *Environment:* the environment surrounding the target industry or service (Bakry, 2004).

## 4.5   OVERVIEW OF ISF

ISF is designed for all types of organizations and entities dealing with information security standards. ISF works by mapping objects or problems into an abstraction group called "domains". The main function of ISF is to assist organizations map the ISO 27001 components, such as refined questions, assessment issues, controls, and clauses into the related domains to obtain the organizations' readiness level and information security capabilities (RISC) investigation. ISF consists of 6 main components identified as domains, namely *organization, stakeholder, tool and technology, policy, culture*, and *knowledge* (Figure 4.1). These domains are associated with the critical components within an organization that influence the organization's business processes that deal with information security.

ISF tries to fill the gap that exists in the 9STAF (Chapter 9 – *State-of-the-Art-Framework*), further discussed in subsection 4.6. Those gaps are measurable quantitative valuation, mathematical modeling, algorithm, and applicability for software development. ISF is designed to help organizations perform ISO 27001 compliance projects efficiently.

Compliance to a standard is a process of comparing the applied controls of an organization with those in the standard. It is basically a gap analysis in which the differences between current circumstances and the standard are discovered. The task of checking conformity level helps an organization to determine its relative position to the standard, which is useful for the certification process.

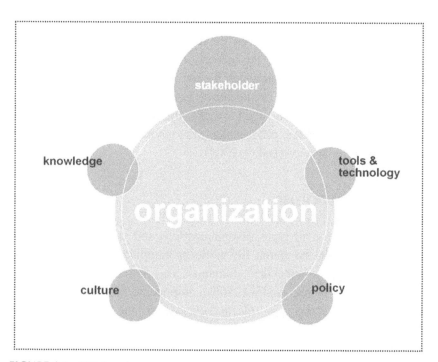

**FIGURE 4.1**   ISF domains.

## 4.5.1   *ISF DOMAINS*

ISF consists of six main components identified as domains, the explanations of each are shown in the subsection below.

### 4.5.1.1   Stakeholder

A person, group, or organization that has direct or indirect stake in an organization that affect or can be affected by the organization's actions, objectives, and policies in achieving their information security objectives.

### 4.5.1.2   Tool and Technology

Things used by an organization to support and enable the business processes in dealing with information security, upon the industry or the service concerned. The purposeful application of information in the design, production and utilization of goods and services and in the organization of human activities are divided into two categories: (1) tangible: blueprints, models, operating manuals, prototypes; and (2) intangible: consultancy, problem-solving, and training methods.

### 4.5.1.3   Organization

A social unit of people systematically structured and managed to meet a need or to pursue collective goals on a continuing basis, refers to the organization performing the information security compliance project.

### 4.5.1.4   Culture

Determines what is acceptable or unacceptable, important or unimportant, right or wrong, workable or unworkable based on information security and ISO 27001 controls. *Organization Culture* is the values and behaviors that contribute to the unique social and psychological environment of an organization, and it is the sum total of an organization's past and current assumptions.

## 4.5.1.5 Policy

Typically described as a principle or rule to guide decisions and achieve rational outcome(s), the information security policy of the organization refers to those necessary to support ISO 27001 certification.

## 4.5.1.6 Knowledge

In an organizational context, knowledge is the sum of what is known and resides in the intelligence and competence of security awareness within organizations. Knowledge has come to be recognized as a factor for information security and ISO 27001 certification.

## 4.6   FEATURES COMPARISON

Table 4.2 shows the comparative study between the 9STAF. The comparison parameters were based on scholarly publications, which also indicated that all those frameworks have three basic features as compulsory components, namely: mapping object(s), abstraction and qualitative measurable. Broader comparisons can be conducted by adding four parameters: (1) quantitative measurable, (2) mathematical formulae/equations, (3) computer algorithms, and (4) software and application. With four additional parameters, only STOPE is the framework in which quantitative measurement and mathematical models are used to support their approach.

Based on these results, ISF is relatively close to STOPE. ISF adopts the abstraction domain approach from STOPE, makes refinements, customizations and extensions for further development. ISF is equipped with a computer algorithm and implementation as software (ISM) to directly assist organizations in measuring RISC level with high accuracy and complete features.

## 4.7   ISF COMPONENTS

### 4.7.1   STRUCTURE

The ISF is composed of the following six levels: *refined question, assessment issue, control, clause, domain*, and *top domain* (Figure 4.2). These

**TABLE 4.2** 9STAF Comparative Parameters

| Framework | 9STAF three-basic features | | | | | | |
|---|---|---|---|---|---|---|---|
| | Mapping Objects | Abstraction | Qualitative Measurable | Quantitative Measurable | Mathematical Formula | Algorithm | Software and Application |
| A Framework for the Governance of Information Security (Posthumus and Solms, 2004) | √ | √ | √ | ○ | ○ | ○ | ○ |
| A Framework for Information Security Management Based on Guiding Standard: A United States Perspective (Sipior and Ward, 2008) | √ | √ | √ | ○ | ○ | ○ | ○ |
| A Security Framework for Information Systems Outsourcing (Fink, 1994) | √ | √ | √ | ○ | ○ | ○ | ○ |
| Information Security Management: A Hierarchical Framework for Various Approaches (Eloff and Solms, 2000) | √ | √ | √ | ○ | ○ | ○ | ○ |
| Information Security Governance Framework (Ohki et al., 2009) | √ | √ | √ | ○ | ○ | ○ | ○ |
| Queensland Government Information Security Policy Framework (QGISPF, 2009) | √ | √ | √ | ○ | ○ | ○ | ○ |
| STOPE Methodology (Bakry, 2004) | √ | √ | √ | √ | √ | ○ | ○ |
| A Security Audit Framework for Security Management in the Enterprise (Onwubiko, 2009) | √ | √ | √ | ○ | ○ | ○ | ○ |
| Multimedia Information Security Architecture Framework (Susanto and Muhaya, 2010) | √ | √ | √ | ○ | ○ | ○ | ○ |
| ISF – Integrated Solution Framework | √ | √ | √ | √ | √ | √ | √ |

√ : *Available/Detected in Scholarly Publication.*

○: *Not Available/Not Detected in Scholarly Publication.*

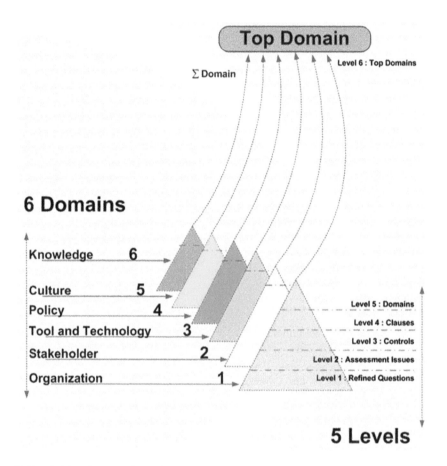

FIGURE 4.2   ISF domains and levels.

levels basically map ISF parameters with the corresponding components of ISO 27001. ISF adopts the bottom-up solution approach, starting with *refined question* as the lowest level. The details of each level are as follows:

1. The first level of ISF is associated with *refined questions* as the basic parameter for measuring ISO 27001 compliance. These parameters are related to infosec circumstances and directly assessed by the users (respondents). A combination of refined questions makes up an assessment issue in ISF (Figure 4.2, level 1).

2. The second level of ISF is *assessment issue* from ISO 27001 to achieve of the information security objectives (Figure 4.2, level 2).

3.  The third level of ISF is related to the sub-parts of the standard, which are concerned with *security control* (Figure 4.2, level 3).

4.  The fourth level is the *clause* or *security objective* organized according to their relationship with each of the ISF domains (Figure 4.2, level 4).

5.  The fifth level is the *domain* as the top layer that has organized within it all the refined questions, assessment issues, controls, and clauses. The score result from this domain is the organization's indicator for readiness level regarding it (Figure 4.2, level 5).

6.  *Top domain* level is the integration of all ISF domains, which indicates the overall status of an organization readiness level (Figure 4.2, level 6).

### 4.7.2 MAPPING ISO 27001 COMPONENTS TO ISF DOMAINS

RISC measurement works on stages by mapping ISO 27001 components (control, clause, and assessment issues) to ISF domains (Figure 4.3). Each domain is handled by the respondent through the user interface to produce a score as the RISC indicator result. An indicator result is a score that describes an organization's circumstances associated with readiness level and information security capabilities with respect to ISO 27001.

For instance, the domain *tools and technology* has 30 parameters of assessment issues. As a result, the performance achieved is 3.75 out of 4. Referring to the scale in ISF, it indicates that the organization's score for RISC investigation is between *above average implemented* and *fully implemented* on respective controls and domain.

ISF provides an efficient way for organizations to perform self-assessment to find out their extent of ISO 27001 compliance. This self-assessment makes it easier for organizations to prepare SoA documents. The certification auditors use this SoA report as the main guideline for the audit. SoA is the central document that defines how an organization will be (or has been) implementing controls, and define which of the suggested 133 controls, including 21 essential controls (Alfantookh, 2010; Bakry, 2009; Susanto et al., 2011) (Table 4.24) of security measures will

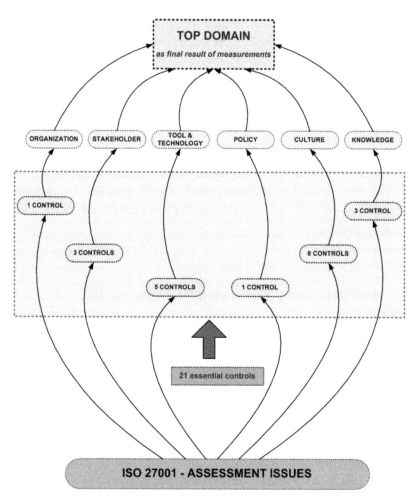

**FIGURE 4.3**   Mapping ISO 27001 components to ISF domains.

apply, and for those that are applicable, the way they will be implemented (Calder, 2006; ISO, 2005).

Each ISF domain is associated with the respective components and related issues of ISO 27001. These issues consists 7 clauses, 21 essential controls, 144 assessment issues and more than 266 refined questions (Bakry, 2004; Susanto et al., 2011a, 2011b, 2012a, 2012b). The main components are described in the following subsections.

## 4.7.2.1 Organization Domain

This domain consists of:
- One clause : organization of information security
- One essential control namely Organization of information security: *allocation of information security responsibilities* (Table 4.3).

## 4.7.2.2 Stakeholder Domain

This domain consists of:
- One clause: Human Resources Security
- Three essential controls:
  i. Human resources security: *management responsibilities* (Table 4.4)
  ii. Human resources security: *Information security awareness, education and training* (Table 4.5)
  iii. Human resources security: *disciplinary process* (Table 4.6)

## 4.7.2.3 Tools and Technology Domain

This domain consists of:
- One clause: Information Systems Acquisition, Development and Maintenance

**TABLE 4.3** Assessing the Control Concerned with the Organization Domain

| *All information security responsibilities should be clearly defined.* | | |
|---|---|---|
| **Assessment Issues** | **Refined Questions** | **Answer** |
| Objects of responsibility | Are the assets and security processes clearly identified? | Levels |
| Levels of responsibility | Does the position of the information security manager that takes the overall responsibility, exist? | Levels |
| | Are the authorization levels identified? | Levels |
| | Does each asset have a person responsible? | Levels |
| | Does each security process have a person responsible? | Levels |
| Documentation | Does the reporting of the above exist? | Levels |

TABLE 4.4 Assessing the Control for the Stakeholder Domain on Management
Responsibilities

| *Management should require employees, contractors and third party users to apply security in accordance with established policies and procedures of the organizations.* | | |
|---|---|---|
| **Assessment Issues** | **Refined Questions** | **Answer** |
| **Responsibilities of employees** | Is the role specified and assigned? | Levels |
| | Is the relevant security awareness achieved? | Levels |
| | Do the security guidelines exist? | Levels |
| | Does conformity with employment conditions exist? | Levels |
| | Does the motivation to fulfill security requirements exist? | Levels |
| | Are the suitable security skills maintained? | Levels |
| **Responsibilities of contractors** | Is the role specified and assigned? | Levels |
| | Is the relevant security awareness achieved? | Levels |
| | Do the security guidelines exist? | Levels |
| | Does conformity with employment conditions exist? | Levels |
| | Does the motivation to fulfill security requirements exist? | Levels |
| | Are the suitable security skills maintained? | Levels |
| **Responsibilities of third party users** | Is the Role specified and assigned? | Levels |
| | Is the relevant security awareness achieved? | Levels |
| | Do the security guidelines exist? | Levels |
| | Does conformity with employment conditions exist? | Levels |
| | Does the motivation to fulfill security requirements exist? | Levels |
| | Are the suitable security skills maintained? | Levels |
| **Documentation** | Does reporting of the above exist? | Levels |

- Five essential controls:
  i. Correct processing in applications: *input data validation* (Table 4.7).
  ii. *Output data validation* (Table 4.8).

**TABLE 4.5**   Assessing the Control for the Stakeholder Domain on *Information Security Awareness, Education and Training*

| *All employees of the organization and, where relevant, contractors and third party users should receive appropriate awareness training and regular updates in organizational policies and procedures relevant to their job function* | | |
|---|---|---|
| **Assessment Issues** | **Refined Questions** | **Answer** |
| **Are awareness, education and training of employees** | Do business policies and procedures exist? | Levels |
| | Do security requirements exist? | Levels |
| | Do legal responsibilities exist? | Levels |
| | Does the correct use of information processing facilities exist? | Levels |
| | Do regular updates exist? | Levels |
| **Are awareness, education and training of contractors** | Do business policies and procedures exist? | Levels |
| | Do security requirements exist? | Levels |
| | Do legal responsibilities exist? | Levels |
| | Does the correct use of information processing facilities exist? | Levels |
| | Do regular updates exist? | Levels |
| **Are awareness, education and training of third party users** | Do business policies and procedures exist? | Levels |
| | Do security requirements exist? | Levels |
| | Do legal responsibilities exist? | Levels |
| | Does the correct use of information processing facilities existing? | Levels |
| | Do regular updates exist? | Levels |
| **Documentation** | Does reporting of the above exist? | Levels |

iii. *Control of internal processing* (Table 4.9).
iv. *Message integrity* (Table 4.10).
v. *Control of technical vulnerabilities* (Table 4.11).

## 4.7.2.4   Policy Domain

This domain consists of:
- One clause: Information Security Policy
- One essential control namely information security policy: *document* (Table 4.12)

**TABLE 4.6**   Assessing the Control for the Stakeholder Domain on *Disciplinary Process*

*There should be a formal disciplinary process for employees who have committed a security breach*

| Assessment Issues | Refined Questions | Answer |
|---|---|---|
| Condition | Has the process commenced only after verification that a security breach occurred? | Levels |
| | Is the correct and fair treatment ensured? | Levels |
| Assessment considerations | Is the impact of breach on business considered? | Levels |
| | Is the nature and gravity of the breach considered? | Levels |
| | Is the repetition of the breach considered? | Levels |
| | Is the received past relevant training considered? | Levels |
| Serious cases considerations | Is the instant denial of access rights considered? | Levels |
| | Is the instant removal of duties considered? | Levels |
| | Is the instant escorting out of site considered? | Levels |
| Rules taken into account | Are the business contracts maintained? | Levels |
| | Are the relevant legislations maintained? | Levels |
| Documentation | Does the reporting of the above exists? | Levels |

**TABLE 4.7**   Assessing the Control for the Tool and Technology Domain on *Input Data Validation*

*Data input to applications should be validated to ensure that this data is correct and appropriate*

| Assessment Issue | Refined Question | Answer |
|---|---|---|
| Existence | Plausibility checks exist to test the output data reasonability? | Levels |
| Validation | Does the examination for the input business transaction, standing data and parameter tables exist? | Levels |
| | Does automatic examination exist? | Levels |
| | Is the periodic review and inspection available? | Levels |
| | Does the response of procedures to validation exist? | Levels |
| Management | Does the logging of events exist? | Levels |
| Accountability | Are the responsibilities defined? | Levels |
| Documentation | Does reporting of the above exist? | Levels |

**TABLE 4.8** Assessing the Control for the Tool and Technology Domain on *Output Data Validation*

| Assessment Issue | Refined Question | Answer |
|---|---|---|
| *Data output from an application should be validated to ensure that the processing of stored information is correct and appropriate to the circumstances* | | |
| Existence | Plausibility checks exist to test the output data reasonability? | Levels |
| Validation | Is the provided information for a reader or subsequent processing system sufficient to determine the accuracy, completeness, precision and classification of the information? | Levels |
| | Does periodic inspection exist? | Levels |
| | Does responding procedures validation test exist? | Levels |
| Practice | Does checking that programs are run in order exist? | Levels |
| | Does checking that programs are run at the correct time exist? | Levels |
| Accountability | Are the responsibilities defined? | Levels |
| Documentation | Does reporting of the above exist? | Levels |

## 4.7.2.5 Culture domain

This domain consists of:

- Two clauses:
  - i. Information Security Incident Management
  - ii. Business Continuity Management
- Eight essential controls:
  - i. Information security incident management: *responsibilities and procedures* (Table 4.13).
  - ii. Information security incident management: *learning from information security incidents* (Table 4.14).
  - iii. Information security incident management: *collection of evidence* (Table 4.15).
  - iv. Business continuity management: *business continuity management process* (Table 4.16).
  - v. Business continuity management: *business continuity and risk assessment* (Table 4.17).

**TABLE 4.9**   Assessing the Control for the Tool and Technology Domain on *Control of Internal Processing*

*Validation checks should be incorporated into applications to detect any corruption of information through processing errors or deliberate acts.*

| Assessment Issue | Refined Question | Answer |
|---|---|---|
| Validation | Does validation of generated data or software, exist? | Levels |
| | Does validation of downloaded data or software, exists? | Levels |
| | Does validation of the uploaded data or software, exist? | Levels |
| Protection | Does the use of programs that provide recovery from failure exists? | Levels |
| | Does the termination of programs at failure exist? | Levels |
| | Does protection against attacks exist? | Levels |
| Practice | Does checking that programs are run in order exist? | Levels |
| | Does checking that programs are run at the correct time exist? | Levels |
| | Does checking that programs terminate correctly exist? | Levels |
| | Does the logging of events exist? | Levels |
| Accountability | Are the responsibilities defined? | Levels |
| Documentation | Does reporting of the above exist? | Levels |

**TABLE 4.10**   Assessing the Control for the Tool and Technology Domain on *Message Integrity*

*Requirements for ensuring authenticity and protecting message integrity in applications should be identified, and appropriate controls identified and implemented*

| Assessment Issue | Refined Question | Answer |
|---|---|---|
| Requirements | Are message integrity requirements specified? | Levels |
| Protection | Are message integrity protection measures implemented? | Levels |
| | Implemented protection measures are suitable to message integrity requirements | Levels |
| Practice | Does the logging of events exist? | Levels |
| Accountability | Are the responsibilities defined? | Levels |
| Documentation | Does reporting of the above exist? | Levels |

vi.   Business continuity management: *developing and implementing continuity plans including information security* (Table 4.18).

**TABLE 4.11**   Assessing the Control for the Tool and Technology Domain on *Control of Technical Vulnerabilities*

| *Timely information about technical vulnerabilities of information systems being used should be obtained, the organization exposure to such vulnerabilities evaluated and appropriate measures taken to address the associated risk* | | |
|---|---|---|
| **Assessment Issue** | **Refined Question** | **Answer** |
| **Inventory of technical assets** | Do the technical specifications of systems and their components exist? | Levels |
| **Vulnerability** | Are the vulnerabilities of technical assets identified? | Levels |
| | Are the risks associated with vulnerabilities identified? | Levels |
| **Protection** | Protection measures that respond to risks are identified | Levels |
| | Are the protection tools evaluated before use? | Levels |
| | Does the awareness on potential vulnerabilities among the right people exist? | Levels |
| **Practice** | Does the monitoring to manage problems exist? | Levels |
| | Does logging of events exist? | Levels |
| **Accountability** | Do the defined responsibilities exist? | Levels |
| **Documentation** | Does reporting of the above exist? | Levels |

    vii.  Business continuity management: *business continuity planning framework* (Table 4.19).

   viii.  Business continuity management: *testing, maintaining and re-assessing business continuity plans* (Table 4.20).

## 4.7.2.6   Knowledge Domain

This domain consists of:

- One clause: Compliance
- Three essential controls:
  - i.  Compliance: *Intellectual property rights* (Table 4.21).
  - ii.  Compliance: *Protection of organizational records* (Table 4.22).
  - iii.  Compliance: *Data protection and privacy of personal information* (Table 4.23).

**TABLE 4.12**   Assessing the Control for the Policy Domain on *Document*

| An information security policy document should be approved by management, published and communicated to all employees and relevant external parties | | |
|---|---|---|
| **Assessment Issue** | **Refined Question** | **Answer** |
| Existence | Is the information security policy document available? | Levels |
| Approval | Is the information security policy document approved by the management? | Levels |
| Publication | Is the information security policy document published? | Levels |
| Internal commu-nication | Is the information security policy document communi-cated to all ICT employees? | Levels |
| | Is the information security policy document communi-cated to all ICT users? | Levels |
| External com-munication | Is the information security policy document communi-cated to relevant external parties? | Levels |
| Documentation | Does reporting of the above exists? | Levels |

**TABLE 4.13**   Assessing the Control for the Culture Domain on *Responsibilities and Procedures*

| Management responsibilities and procedures should be established to ensure a quick, effective, and orderly response to information security incidents | | |
|---|---|---|
| **Assessment Issue** | **Refined Question** | **Answer** |
| Responsibility | Are the procedures concerned with information security incidents approved by management? | Levels |
| | The commitment level to the procedures concerned with security incidents is? | Levels |
| Types of incidents | Are management procedures established for each key incident? | Levels |
| | Are all key incidents identified? | Levels |
| Producers | Do the Incident procedures consider business integrity? | Levels |

## 4.8   THE MATHEMATICAL NOTATION

To counting RISC score, we developed a mathematical notation (formula) as application of the ISF approach. This formula has four levels of iteration counts. The lowest level deals with the assessment issues. The

**TABLE 4.14**   Assessing the Control for the Culture Domain on *Learning from Information Security Incidents*

| There should be mechanisms in place to enable the types, volumes, and costs of information security incidents to be quantified and monitored | | |
| --- | --- | --- |
| **Assessment Issue** | **Refined Question** | **Answer** |
| **Learning about incidents** | Are the incidents assessed by type? | Levels |
| | Are the incidents assessed by occurrence? | Levels |
| | Are the incidents assessed by impact? | Levels |
| | Are the incidents are assessed by cost? | Levels |
| **Conclusions** | Are the special control responses developed? | Levels |
| | Are the developed responses implemented? | Levels |
| **Documentation** | Does reporting of the above exist? | Levels |
| | Are the incident procedures based on full understanding of the nature of incidents? | Levels |
| | Are the incident procedures providing suitable solutions to the incidents? | Levels |
| | Are the Incident procedures providing authority only to trusted personnel? | Levels |
| **Contingency plans** | Are the contingency plans prepared for the key incidents that require such plans? | Levels |
| **Audit trail** | Are the causes of incidents trailed internally? | Levels |

iterative processes are conducted until the highest level, top domains that demonstrate readiness level and information security capabilities toward ISO 27001 compliance.

Several variables were constructed to indicate parameters of ISF components, such as *l* as *assessment issue, k* as *control, j* as *clause, i* as *domain* and *h* as *top domain.*

### 4.8.1  CONTROL

The score for Control is the total summation of the assessment issue(s), divided by the number of assessment issue(s), *n-assessment issue(s)*. The above can be expressed in the following formula and the notation could be described as:

**TABLE 4.15**   Assessing the Control for the Culture Domain on *Collection of Evidence*

| *Where a follow-up against a person or organization, after an information security incident involves legal action (either civil or criminal), evidence should be collected, retained, and presented to conform to the rules for evidence laid down in relevant jurisdictions* | | |
| --- | --- | --- |
| **Assessment Issue** | **Refined Question** | **Answer** |
| Understanding the legal system | Is the admissibility of evidence in the legal system understood? | Levels |
| | Is the weight of evidence in the legal system understood? | Levels |
| | Are the evidence requirements identified for internal problems? | Levels |
| | Are the evidence requirements identified for external problems? | Levels |
| | Do the internal procedures consider evidence requirements for possible problems? | Levels |
| | Do the external procedures consider evidence requirements for possible problems? | Levels |
| **Documentation** | Does reporting of the above exist? | Levels |

$$x_j = \sum_{k=1}^{n} \frac{[assessment\ issue]_k}{n} \qquad (1)$$
$$x_j{:}control$$

where (a) $x_j$, (b) $x_j$ is the score of ISO 27001's control derived from the total of assessment issue(s).

### 4.8.2   CLAUSE

The score for clause is the total summation of the control(s), divided by the number of control(s), *n-control(s)*. The notation could be described as;

$$x_i = \sum_{j=1}^{n} \frac{[control]_j}{n} \qquad (2)$$
$$x_i{:}clause$$

**TABLE 4.16** Assessing the Control for the Culture Domain on *Business Continuity Management Process*

| Assessment Issue | Refined Question | Answer |
|---|---|---|
| Identification of critical business processes | Are the critical business processes identified? | Levels |
| | Are the critical business processes prioritized? | Levels |
| | Are the assets associated with critical business processes identified? | Levels |
| Identification of risk | Are the risks identified? | Levels |
| | Are the probabilities of risks identified? | Levels |
| | Are the impacts of risks on business identified? | Levels |
| | Are the risks (incidents) classified according to their level of seriousness? | Levels |
| Protection considerations | Are the information security requirements identified? | Levels |
| | Is the protection of personnel identified? | Levels |
| | Is the protection of processing facilities identified? | Levels |
| | Is the protection of organizational property identified? | Levels |
| | Is purchasing insurance considered? | Levels |
| Procedures | Do the procedures considering the above exist? | Levels |
| | Are the procedures updated regularly? | Levels |
| | Are the procedures incorporated in the business process? | Levels |
| Responsibilities | Are the business continuity responsibilities assigned at the appropriate levels? | Levels |
| Documentation | Does reporting of the above exist? | Levels |

where (a) $x_j$, (b) $x_j$ indicates the score for the clause of ISO 27001 which results from the total of control(s).

## 4.8.3 DOMAIN

The third step is concerned with the investigation of compliance with one ISF domain. The evaluation of these indicators depends on the evaluation of the indicators of compliance of ISO main parts (controls and clause).

**TABLE 4.17**   Assessing the Control for the Culture Domain on *Business Continuity and Risk Assessment*

| Events that can cause interruptions to business processes should be identified, along with the probability and impact of such interruptions and their consequences for information security | | |
|---|---|---|
| **Assessment Issue** | **Refined Question** | **Answer** |
| **The organization** | Are the organization objectives considered? | Levels |
| | Are the organization priorities and criteria considered? | Levels |
| | Are the allowable outage times identified? | Levels |
| | Are the critical resources identified? | Levels |
| **Interruption events** | Are the events (i.e., equipment frailer, human errors and theft) that cause interruption identified? | Levels |
| | Are the probabilities of events identified? | Levels |
| | Are the interruption impacts of events identified? | Levels |
| | Are the recovery periods identified? | Levels |
| | Are the priority risks identified? | Levels |
| **Business continuity plan** | Are the information security requirements identified? | Levels |
| | Is the protection of personnel identified? | Levels |
| | Is the protection of processing facilities identified? | Levels |
| **Documentation** | Does reporting of the above exist? | Levels |

Domain is total summation of the clause(s), divided by the number of clause(s), *n-clause(s)*. The notation could be described as:

$$x_g = \sum_{i=1}^{n} \frac{[domain]_h}{n} \tag{3}$$

$$x_h:domain$$

where (a) $x_h$, (b) $x_h$ declares the grade for the domain which results from the total of the clauses.

## 4.8.4   TOP DOMAIN

Top domain is the final step, and it is the overall indicator of all ISF domains. It basically tells the compliance level to ISO 27001. The top domain score is based on the six ISF domains.

**TABLE 4.18**   Assessing the Control for the Culture Domain on *Developing and Implementing Continuity Plans Including Information Security*

| *Plans should be developed and implemented to maintain and restore operations and ensure availability of information at the required level and in the required time scales following interruption to, or failure of, critical business processes* | | |
| --- | --- | --- |
| **Assessment Issue** | **Refined Question** | **Answer** |
| **Requirements** | Is the focus made on business objectives? | Levels |
| | Are the organizational vulnerabilities identified? | Levels |
| | Are the required resources identified? | Levels |
| | Are the required services identified? | Levels |
| | Are the education and training of staff adequate? | Levels |
| **Planning process requirements** | Are the responsibilities assigned at different levels? | Levels |
| | Do the recovery procedures exist? | Levels |
| | Are the business dependencies notified? | Levels |
| | Is the plan incorporated in the business processes? | Levels |
| | Is the testing and updating of the plan performed? | Levels |
| **Documentation** | Is the business continuity plan considering all the above documented? | Levels |
| | Are the copies of the plan are stored in remote locations? | Levels |

$$x_g = \sum_{i=1}^{n} \frac{[domain]_i}{n} \tag{4}$$

$x_g$:top domain

where (a) $x_g$ (b) $x_g$ denotes the score of the top domain which results from the total of ISF's six domains.

### 4.8.5   SUBSTITUTION TO SINGLE EQUATION

To simplify the above formula; *control [$x_j$], clause [$x_j$], domain [$x_h$]*, and, *top domain [$x_g$]*, could be substituted to become a single mathematical equation. The substitution steps of *(a), (b), (c)* and *(d)* to the comprehensive notation in a single mathematical equation is shown as follows:

**TABLE 4.19**    Assessing the Control for the Culture Domain on *Business Continuity Planning Framework*

| *A single framework of business continuity plans should be maintained to ensure all plans are consistent to consistently address information security requirements, and to identify priorities for testing and maintenance* | | |
|---|---|---|
| **Assessment Issue** | **Refined Question** | **Answer** |
| **Continuity approach** | Is the approach for continuity (e.g., information availability and security) identified? | Levels |
| | Are the conditions of activation of plans identified? | Levels |
| | Are the responsibilities for action specified? | Levels |
| | Are the owners of plans identified? | Levels |
| **Planning process requirements** | Do the emergency procedures exist? | Levels |
| | Do the manual fallback plans exist? | Levels |
| | Do the temporary procedures exist? | Levels |
| | Does the resumption plan exist? | Levels |
| | Does the maintenance plan exist? | Levels |
| | Does the awareness plan exist? | Levels |
| | Does the education and training plan exist? | Levels |
| | Are the responsibilities of individuals assigned? | Levels |
| | Do the assets and resources for the work exist? | Levels |
| **Documentation** | Does reporting of the above exist? | Levels |

$$x_g = \sum_{i=1}^{n} \frac{[domain]_i}{n}$$

$$x_g = \sum_{i=1}^{n} \frac{[b]_i}{n}$$

$$x_g = \sum_{i=1}^{n} \frac{\left[\sum_{j=1}^{n} \frac{[clause]_j}{n}\right]_i}{n}$$

**TABLE 4.20** Assessing the Control for the Culture Domain on Testing, Maintaining and Re-Assessing Business Continuity Plans

| *Business continuity plans should be tested and updated regularly to ensure that they are up to date and effective* | | |
|---|---|---|
| **Assessment Issue** | **Refined Question** | **Answer** |
| **Recovery team** | Are all (concerned) aware of the plans? | Levels |
| | Are the responsibilities for business continuity assigned? | Levels |
| | Are the responsibilities for information security assigned? | Levels |
| | Are the responsibilities for regular reviews assigned? | Levels |
| | Are the roles at emergencies clear to all? | Levels |
| **Assurance of plans** | Do the scenarios (interruptions) exist? | Levels |
| | Is the simulation performed to train people? | Levels |
| | Is the restoration of information systems tested? | Levels |
| | Are the alternative sites recovery operations tested? | Levels |
| | Are the suppliers' facilities and services tested? | Levels |
| | Are complete rehearsals performed? | Levels |
| **Updating considerations** | Is the change in personnel updated? | Levels |
| | Is the change in contact information updated? | Levels |
| | Is the change in business strategy updated? | Levels |
| | Is the change in resources updated? | Levels |
| | Is the change in legislation updated? | Levels |
| | Is the change in suppliers and customers updated? | Levels |
| | Is change in processes updated? | Levels |
| | Is change in risk updated? | Levels |
| **Documentation** | Does reporting of the above exist? | Levels |

$$x_g = \sum_{i=1}^{n} \frac{\left[ \sum_{j=1}^{n} \frac{[a]_j}{n} \right]_i}{n}$$

Therefore, for the six domains of ISF (Organization, Stakeholder, Tool and Technology, Knowledge, Culture, and Policy), Eq. ($x_g$) can be expressed using the formula:

**TABLE 4.21** Assessing the Control for the Knowledge Domain on *Intellectual Property Rights*

| Appropriate procedures should be implemented to ensure compliance with legislative, regulatory, and contractual requirements on the use of material in respect of which there may be intellectual property rights and on the use of proprietary software products | | |
|---|---|---|
| **Assessment Issue** | **Refined Question** | **Answer** |
| Understanding intellectual property rights (IPR) | Is the document, software and design copyright considered? | Levels |
| | Is the source code licenses considered? | Levels |
| | Is the patents considered? | Levels |
| | Is the trademarks considered? | Levels |
| IPR policy | Does the intellectual property rights (IPR) policy exist? | Levels |
| | Does the IPR policy awareness have a wide reach? | Levels |
| | Are all partners trusted in IPR? | Levels |
| | Are the IPR legal requirements observed? | Levels |
| | Are IPR contractual requirements observed? | Levels |
| Documentation | Does reporting of the above exist? | Levels |

$$x_g = \sum_{i=1}^{n} \frac{\left[\sum_{j=1}^{n} \frac{\left[\sum_{k=1}^{n} \frac{[control]_k}{n}\right]_j}{n}\right]_i}{n} \tag{5}$$

$$x^a = \sum_{\varrho}^{I} \frac{\varrho}{\left[\sum_{\mathrm{s}\mathrm{f}}^{\mathrm{f}=I} \frac{\mathrm{s}\mathrm{f}}{\left[\sum_{\mathrm{s}\mathrm{f}}^{\backslash=I} \frac{\mathrm{s}\mathrm{f}}{\left[\sum_{\mathrm{s}\mathrm{f}}^{\mathrm{K}=I} \frac{\mathrm{s}\mathrm{f}}{[control]^{\kappa}}\right]^{\backslash}}\right]^{\mathrm{f}}}\right]} \tag{6}$$

## 4.9  COMPUTER ALGORITHM

This computer algorithm is derived from the mathematical formula mentioned, implementing a step-by-step procedure for the investigation of

**TABLE 4.22**   Assessing the Control for the Knowledge Domain on *Protection of Organizational Records*

| *Important records should be protected from loss, destruction, and falsification, in accordance with salutary, regulatory, contractual, and business requirements* | | |
|---|---|---|
| **Assessment Issue** | **Refined Question** | **Answer** |
| **Understanding organizational records** | Do accounting records exist? | Levels |
| | Do the relevant database records exist? | Levels |
| | Are the transaction and audit logs available? | Levels |
| | Do the operational procedures exist? | Levels |
| | Do cryptographic keying materials exist? | Levels |
| | Are sources of key information maintained? | Levels |
| **Record specification** | Are the retention periods of records identified? | Levels |
| | Are the storage media of records identified? | Levels |
| | Are the validities of the storage media identified? | Levels |
| | Are the disposal rules of records identified? | Levels |
| **Control of records** | Are the records protected from information loss? | Levels |
| | Are the records protected from information destruction? | Levels |
| | Are the records protected from information falsification? | Levels |
| **Documentation** | Does reporting of the above exist? | Levels |

each domain. This algorithm describes the ISF work-process mechanism as computer logic from a programming point of view. It calculates the six domains of the framework in a hierarchical approach from lowest level of framework which are the assessment issues, then clauses, domains and the top domain. The measurement and investigation result indicates an organization's readiness for ISO 27001 compliance.

### 4.9.1   CONTROL

This investigation stage is concerned with the achievement of ISF security controls. The evaluation of these indicators (controls) is based on the evaluation of the previous indicator level. In this case, it is associated with the assessment issues within the control.

**TABLE 4.23**   Assessing the Control for the Knowledge Domain on Data Protection and
Privacy of Personal Information

| Data protection and privacy should be ensured as required in relevant legislation, regulations, and if applicable, contractual clauses | | |
| --- | --- | --- |
| **Assessment Issue** | **Refined Question** | **Answer** |
| **Legislations on: collection, processing and transmission of personal information** | Do the country level legislations exist: Information privacy policy available? | Levels |
| | Does the organization level privacy policy exist? | Levels |
| | Do the technical protections measures exist? | Levels |
| | Do the protection procedures exist? | Levels |
| | Are the legislations observed? | Levels |
| **Responsibilities and practices** | Is the data protection officer appointed? | Levels |
| | Are the managers' responsibilities and guidelines identified and applied? | Levels |
| | Are the user's responsibilities and guidelines identified and applied? | Levels |
| | Are the service providers' responsibilities and guidelines identified and applied? | Levels |

## High-level description:

1. Determine that the each control has numbers of assessment issue(s).
2. Sum all assessment issue values and divide by the number of assessment issues.
3. The score indicates the RISC measurement for the control.

## Low-level description:

*# Algorithm level-4 for measuring assessment issues # (Figure 4.4)*
*Initiate,*
*Step 1 [measuring assessment issues]*
*For $k <> 0$ then*
   *$j = k + (k+1)$*
         *Else $Xj = j/n$*
      *End;*
*AssessmentIssue#1, AssessmentIssue #2, ..., AssessmentIssue#n as*
*property of dependent Clause;*

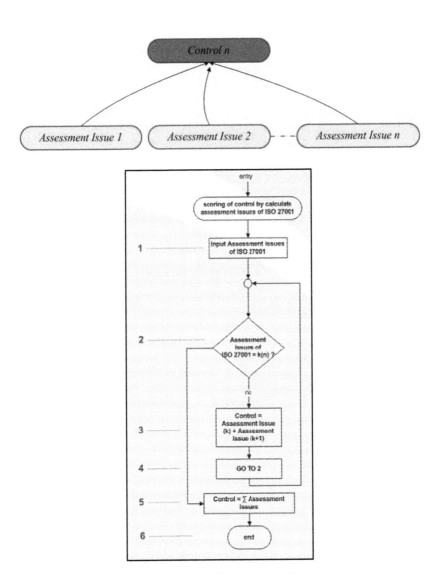

**FIGURE 4.4**   Algorithm level-4 for measuring assessment issues.

$TotalAssessmentIssue = AssessmentIssue \#1 + AssessmentIssue \#2 +$
$... + AssessmentIssue \#n;$
$ValControl\#1 = TotalAssessmentIssue/n;$
$Control\#1 = ValControl\#1$
$\# end \#$

## 4.9.2   CLAUSE

This investigation stage is concerned with the achievement of ISF security clauses. The evaluation of clause (objective) level is based on the evaluation of the previous indicators level. In this case, it is associated with the controls concerned.

**High-level description:**
1. Assume that the each clause has numbers of security control(s).
2. Sum all security controls and divide by the number of controls.
3. The score indicates the RISC measurement for the clause.

**Low-level description:**
*# Algorithm level-3 for measuring assessment control # (Figure 4.5)*
*Initiate;*
*Step 2 [measuring control]*
*For j <> 0 then*
      *i = j + (j+1)*
              *Else Xi = i/n*
      *End;*
*Coltrol#1, Control#2, ...,Control#n as property of dependent Clause;*
*TotalValControl = ValControl#1 + ValControl#2 + ... + ValControl#n;*
*ValClause#1 =TotalValControl/n;*
*Clause#1 = ValClause#1*
*# end #*

## 4.9.3   DOMAIN

This investigation stage is concerned with the evaluation of ISF domains. The evaluation of these domains is based on the evaluation of the previous level of indicators, namely the clauses (objectives).

**High-level description:**
1. Assume that the each domain has numbers of clause(s).
2. Sum all clauses and divide by the number of clauses.
3. The score indicates the RISC measurement for the domain.

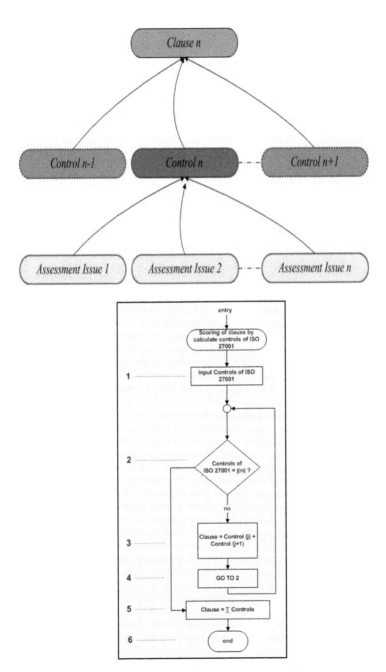

**FIGURE 4.5**   Algorithm level-3 for measuring assessment control.

TABLE 4.24 A View of ISO 27001 Domain, Clauses, Controls and Essential Controls

| Domain | Structure of ISO 27001 | | | |
|---|---|---|---|---|
| | Clause(s) | | Controls | |
| | | Total | Essential | |
| | | | Section | No. |
| Policy | Information Security Policy | 2 | 5.1.1 | 1 |
| Tool and Technology | Information Systems Acquisition, Development, and Maintenance | 16 | 12.2.1 | 5 |
| | | | 12.2.2 | |
| | | | 12.2.3 | |
| | | | 12.2.4 | |
| | | | 12.6.1 | |
| Organization | Organization of Information Security | 11 | 6.1.3 | 1 |
| | Information Security Incident Management | 5 | 13.2.1 | 3 |
| | | | 13.2.2 | |
| | | | 13.2.3 | |
| Culture | Business Continuity Management | 5 | 14.1.1 | 5 |
| | | | 14.1.2 | |
| | | | 14.1.3 | |
| | | | 14.1.4 | |
| | | | 14.1.5 | |
| Stakeholder | Human Resources Security | 9 | 8.2.1 | 3 |
| | | | 8.2.2 | |
| | | | 8.2.3 | |
| Knowledge | Compliance | 10 | 15.1.2 | 3 |
| | | | 15.1.3 | |
| | | | 15.1.4 | |
| Total controls | | 133 | | 21 |

## Low-level description:

*# Algorithm level-2 for measuring assessment clause # (Figure 4.6)*
*Initiate;*
*Step 3 [measuring clause]*
*For i <> 0 then*
    *h = i + (i+1)*
      *Else Xh = h/n*

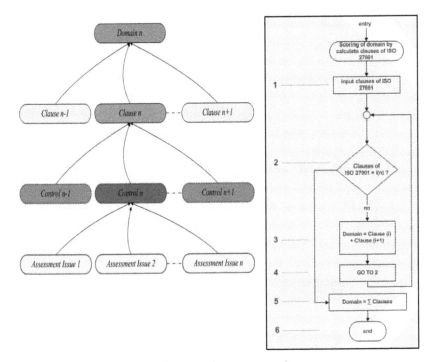

**FIGURE 4.6**   Algorithm level-4 for measuring assessment issues.

*End;*
*Clause #1, Clause #2, ..., Clause #n as property of dependent Domain;*
*TotalValClause = ValClause#1 + ValClause#2 + ... + ValClause#n;*
*ValDomain#1 = TotalValClause/n;*
*Domain#1 = ValClause#1*
*# end #*

### 4.9.4   TOP DOMAIN

This investigation stage is concerned with the evaluation of the top domain. The evaluation of this level is correlated with the evaluation of the domains level. This section indicates the assessment of the six domains concerned.

**High-level description:**
   1.   Sum all six domains and divide by 6.

2.  The score indicates the RISC measurement for the organization level.

**Low-level description:**
*# Algorithm level-1 for measuring assessment domain # (Figure 4.7)*
*Initiate;*
*Step 4 [TopDomain]*
*For h <> 0 then*
  $g = h + (h+1)$
    *Else Xg = h/n*
*End;*
*Domain1 #1, Domain #2,..., Domain #6 as property of dependent TopDomain;*

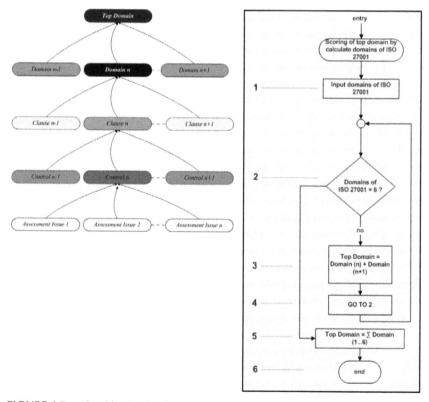

**FIGURE 4.7**    Algorithm level-1 for measuring assessment domain.

$$TotalValDomain = ValDomain\#1 + ValDomain\#2 + ... + ValClause\#6;$$
$$ValTopDomain\#1 = TotalValDomain/6;$$
$$TopDomain = ValTopDomain$$
$$\# \, end \, \#$$

### 4.9.5 HOW ISF FORMULA AND ALGORITHM WORK FOR RISC INVESTIGATION

The output of the investigation process is an indicator that provides an overall ISO 27001 information security status or RISC investigation level for an organization. The RISC investigation has main features as described below:

1. It provides an indicator for each of the ISF domains and an integrated indicator for all the domains (as illustrated in Figure 4.2).
2. It uses a bottom-up approach to calculate indicators starting at the bottom level and gradually moving from one level to another, where the evaluation of each of the higher levels is based on the evaluation of the previous level.
3. The final result, as indicated in the top domain level is the integration of the indicators from the lowest level (assessment issues) to the highest level (domain).

Figure 4.2 and the above model shows the practical stages of the ISF formula and algorithm for RISC investigation. These investigations would produce indicators of compliance at the various levels of the ISF structure. This would help in diagnosing the strengths and the weaknesses of the information security management in the organizations concerned; it would also help direct their efforts toward the issues that require improvements. The RISC investigations process evaluates each of the indicators and calculates scores for the upper level indicator. It refines the investigation of compliance into the following five main steps:

1. The first step investigates indicators at the bottom level, which are the measures associated with the evaluation of ISO 27001's assessment issues.
2. The second step moves up to the investigation of ISO security controls. The evaluation of these indicators will influence that of the clauses (objectives) within which the controls are contained.

Relative weights (in this case the weight is determined by "1") are taken into account, with respect to the relationships of the security controls with their related clauses (objectives).

3.  The third step is investigation of compliance with the ISO clauses (objectives). The evaluation of these indicators depends on the evaluation of the ISO controls. Relative weights (in this case the weight is determine by "1") are also taken into account, with respect to the relationships of the clauses with their parent domain.

4.  The fourth step is the investigation of compliance with one ISF domain. The evaluation of these indicators correlates with the evaluation of the indicators of ISO 27001 compliance.

5.  The fifth step is the final step. It is concerned with the overall indicator of all ISF domains put together collectively, giving the overall RISC measurement. The evaluation of the top domain is based on the evaluation the indicators of compliance of the six ISF domains.

## KEYWORDS

- **ISF domains**
- **ISO**
- **RISC investigation**
- **STOPE**

# CHAPTER 5

# SOFTWARE DEVELOPMENT

## CONTENTS

## 5.1   INTRODUCTION

Software development is the process of developing a software product. The term *software development* includes all activities involved between the conception of the desired software to the final manifestation of the software in a planned and structured process (Jackson, 2001; Wolfgang, 1994). Software development may include new development, prototyping, modification, reuse, re-engineering, maintenance, or any other activities that result in software products (Cockburn & Highsmith, 2001; Highsmith & Cockburn, 2001; Martin, 2003).

Software can be developed for a variety of purposes, the three most common ones are to meet the specific needs of a specific user, to satisfy a perceived need of some set of potential users, or for personal use (e.g., a scientist may write software to automate a task). The need for better quality control for the software development process has given rise to the discipline of software engineering, which aims to apply the systematic approach exemplified in the engineering paradigm to the process of software development. Wahono (2006) mentioned that the software quality (SQ) or software performance (SP) is the theme of research in the history of the science of software engineering. From the product point of view, measuring SQ/SP can be conducted using the McCall taxonomy as the best practice of software measurement (McCall, 1977). Herbsleb (1997) and Gan (2006) stated that a series of parameters for performances indicator were defined, such as advantages, accuracy, precision, stability, ease of operation and graphical user interface.

This chapter consists of the software development stages for integrated solution modeling software (ISM), the software implementation of the framework ISF. ISM provides a user interface to bridge between the user and the RISC measurement approach. ISM has two main functions: security assessment management (SAM/e-Assessment) and security monitoring management (SMM/e-Monitoring), the details of each these functions and the architecture of the software will be discussed shortly.

This chapter aims to answer the research question associated with technical terms and technical barriers, which caused most ISO 27001 implementation projects to be delayed. Technical barriers are one of the biggest obstacles in understanding the ISO 27001 components. If organizations work with consultants for ISO 27001 compliance, then other barriers may arise such as misunderstandings and misperceptions. Consequently, the certification process may take a very long time and be costly (Kosutic, 2010, 2013).

The remaining part of this chapter discusses the specific techniques for software development in more detail, the existing features in the software, the RISC investigation module and the real-time monitoring module.

## 5.2  ISM TECHNOLOGIES AND FRONT-END ARCHITECTURE

The advancement of the client-server and multiuser software development technology affects the ISM development style. With architecture and design that supports a client-server configuration, ISM is expected to be the center point for RISC investigation and network real-time monitoring (Figure 5.1).

ISM has a multiuser database that can be stored in a cloud network. The total capacity accommodated by the database is approximately 50

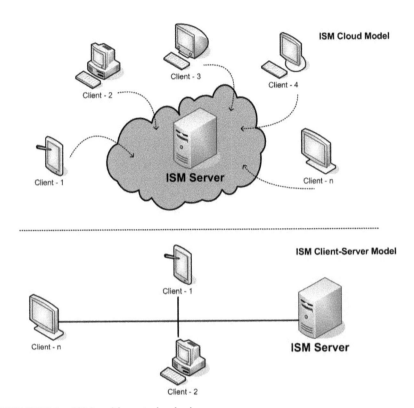

**FIGURE 5.1**    ISM multiuser technologies.

gigabytes. It indicates that the storage and data backup could be performed continuously in a large capacity. In case the database is full, extension through storage replication is available. As a result ISM and its database can be used anytime-anywhere.

Visual Basic (VB) version 6 is carefully chosen to develop the software rather than later versions of VB such as VB.NET, as VB 6 provides modules to support scripting processes. There are several reasons as to why VB 6 is chosen for ISM development: First, VB is not only a language but primarily it is an integrated, interactive development environment (IDE). Second, the VB-IDE has been highly optimized as an application program to support rapid application development (RAD). It is particularly easy to develop graphical user interfaces and to connect them to handler functions provided by the application. Third, the graphical user interface of the VB-IDE provides intuitively appealing views for the management of the program structure in the large and the various types of entities (classes, modules, procedures, forms). Fourth, VB is a component integration language which is attuned to component object model (COM). Fifth, COM components can be written in different languages and then integrated using VB. Sixth, interfaces of COM components can be easily called remotely through distributed COM (DCOM), which makes it easy to construct distributed applications. Seventh, COM components can be embedded in/linked to the application's user interface and also in/to stored documents (Object Linking and Embedding – OLE, Compound Documents). Eighth, VB can support the assembly based or low-level programming, which is useful to track the use of memory resources, the traffic of the file system, and network communication. For instance in the SMM[1] feature of ISM (monitoring), VB allows detection of the connection of computer processes, open ports, process management, suspected denial of services and malicious software (Table 5.1).

The objective of the ISM front-end architecture is to create a clear, complete, easy, and user-friendly interface. This architecture can be classified into three broad ranges of services, which are the RISC investigation module, the network monitoring module, and respondent-personality module.

Respondents, as the organization's representatives, are incorporated in the respondent-personality module, while the RISC investigation module functions as the ISO 27001 compliance measurement and the network monitoring module refers to the real-time network monitoring for suspected information security breaches.

The implementation of ISM proposes to provide a foundation to build an integrated solution that can incorporate organization enhancement processes electronically and paperless for information security self-assessment. The major benefits and reasons for using ISM are having an integrated solution for managing risk, system availability, gain customer trust, reduce compliance cost, and shorten time for preparation for ISO 27001 assessment.

---

[1] Security Monitoring Management: monitors real-time system activity, firewall and network management to provide security monitoring, and potential suspect of security breaches. ISM tools collect event data in real time in to enable immediate analysis and response.

**TABLE 5.1**    Port Detection

```
For n = 0 To 5000
    If Cancel = True Then Command1.Tag = i: Exit Sub
        If sckScan(n).State = sckClosed Then
sckScan(n).Connect strHost, i
DoEvents
GoTonextport
        End If
    Next n
nextport:
lngScanned = lngScanned + 1
ProgressBar1.Value = i
lblStatus.Caption = i & "/" & txtPort2.Text
DoEvents
Next i
tmrScan.Enabled = False
Exit Sub
End Sub


If Option1.Value = True Then
lstPorts.AddItem "Scanning your computer"
Else
lstPorts.AddItem "Scanning" & txtHost.Text
End If
lstPorts.AddItem "From ports" & txtPort1 & "to" & txtPort2
lstPorts.AddItem "Scan started at" & Time & "on" & Date
lstPorts.AddItem "------------------------------------------"
Scan
End Sub


End Sub
Private Sub Option1_Click()
txtHost.Enabled = False
txtHost.Text = "###.###.###.### -or- www.domain.com"
End Sub
```

## 5.3   THE DATABASE

The ISM database is divided into three major entities, namely the user, the ISO 27001 controls and the RISC investigation result. The entity/table for users' access authority represents the users' specification to access the ISM, such as guest, user member, and admin. The relationship type between user and controls is one to many. It indicates that each user can conduct the RISC investigation more than once, which allows the trial and learning assessment (Figure 5.2).

For further analysis on RISC result, this database can also record user titles associated with job position, for example, IT manager, IT staff, information security officer, board of management, expert or consultant. Users can update, delete, and retrieve data from the database for a certain query operation.

Communication link is made between users, interface and the database by using structure query language (SQL), and ADODB[2] connection driver, that serves to perform a series of operations such as select, update, delete, or retrieve records with certain criteria defined by the user. SQL is supported by object linking and embedding (OLE). OLE allows embedding and linking to documents and other objects. For developers, it provides OLE control extension (OCX), a way to develop and use custom user interface elements. Data object enables the transfer of data, and notification of data and database changes. It must be implemented by objects that are supported or embedded in a containing document (Tables 5.2 and 5.3).

## 5.4   SOFTWARE DEVELOPMENT STAGES

There are two essential steps to develop software, regardless of its size and complexity (Boehm, 2000, 2001; Cockburn and Highsmith, 2001). First is analysis, followed by coding (Figure 5.3). This simple implementation approach is required if the effort is sufficiently small and if the final product is to be operated by those who built it for an internal use.

---

[2] ADODB is a database abstraction library for PHP and Python based on the same concept as ActiveX Data Objects. It allows developers to write applications in a fairly consistent way regardless of the underlying database system storing the information. The advantage is that the database system can be changed without re-writing every call to it in the application.

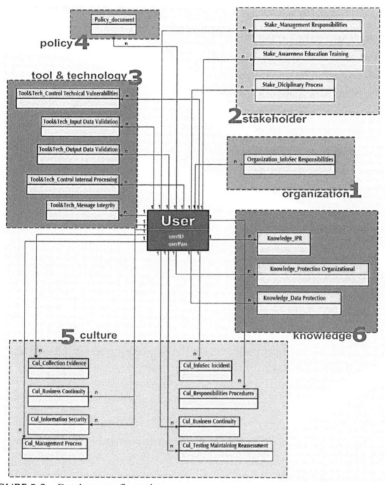

**FIGURE 5.2** Database configuration.

**TABLE 5.2** Database Connection

### Database connection
Dim con As New ADODB.Connection
  Dim rs As New ADODB.Recordset
  Dim flag As Boolean
con.Open ("provider=microsoft.jet.oledb.4.0;data source=" &
App.Path&\ISO27001.mdb")
### Database Query for Technology Input Data Validation Attributes
rs.Open
"select * from TechnologyInputDataValidation", con, 1, 3
rs.MoveFirst

**TABLE 5.3**   ADODB and Recordset Connection Method

### Database Query for Top Stakeholder Domain

Dim con As New ADODB.Connection

Dim rs1 As New ADODB.Recordset

Dim rs2 As New ADODB.Recordset

Dim rs3 As New ADODB.Recordset

Dim rs4 As New ADODB.Recordset

Dim rs5 As New ADODB.Recordset

Dim total

Dim userId

userId = Username

con.Open ("provider=microsoft.jet.oledb.4.0;data source=" &App.Path& "\ISO27001.mdb")

rs1.Open "select TotalScorefromStakeholderMR where username='" &userId& "' order by id desc", con, 1, 3

total1 = rs1.Fields.Item(0)

   txt1.Text = total1

   rs1.Close

rs2.Open "select TotalScorefromStakeholderISAET where username='" &userId& "' order by id desc", con, 1, 3

total2 = rs2.Fields.Item(0)

   txt2.Text = total2

   rs2.Close

rs3.Open "select TotalScorefromStakeholderDP where username='" &userId& "' order by id desc", con, 1, 3

total3 = rs3.Fields.Item(0)

   txt3.Text = total3

   rs3.Close

con.Close

   txt6.Text = (total1 + total2 + total3) / 3

End Sub

A more comprehensive approach to software development is illustrated in Figure 5.4. This approach is called the waterfall approach (WFA) or

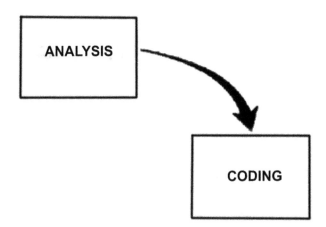

**FIGURE 5.3**   Two essential steps of software development.

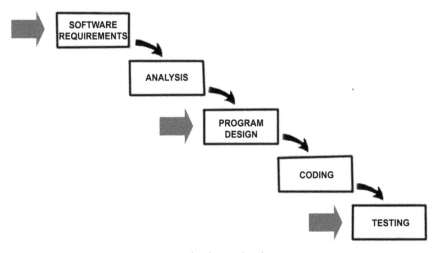

**FIGURE 5.4**   Comprehensive steps of software development

waterfall software process (WSP[3]). The analysis and coding steps are still in the picture, but they are preceded by *requirements analysis*, separated

[3] WSP views the optimal process for software development as a linear or sequential series of phases that takes the developers from initial high-level requirements through system testing and evaluation (Cusumano and Smith, 1995).

WSP begins by writing a specification that is as complete as possible. Next, divide the specification into modules in a more detailed design phase. This process of integration and system testing requires reworking the modules and writing new code to correct problems in the operation or interactions of the pieces due to unforeseen problems as well as mistakes, misconnections, or changes that have crept into the design of the parts during the project (Cusumano and Smith, 1995; MacCormack, 2003).

by a *program design* step and followed by a *testing step* (arrows). Analysis and coding must be planned and staffed differently for best utilization of program resources.

The implementation and software development methodology for ISM is a composite between WSP and SDA[4] (spiral development approach) that combines elements of design in stages. The SP focuses on risk assessment on minimizing implementation risk by breaking a project into smaller functions, procedures and modules. It provides better ease-of-change during the development process, as well as the opportunity to evaluate risks and weigh consideration of project continuation throughout the life cycle. Each cycle involves a progression through the same sequence of steps for each part of the product and for each of its levels of elaboration, from an overall concept-of-operation document down to the coding of each individual program. Each trip around the spiral traverses four basic quadrants which represent the four fundamentals of software development: (1) requirement; (2) analysis; (3) design and implementation; (4) testing and evaluation. The cycle begins with an identification of potential users and their conditions, and ends with review and commitment for evaluation (Figure 5.5).

## 5.4.1   REQUIREMENT STAGE

ISM's development process starts by interviewing and discussing with respondents as potential users to find out their expectations and pretentions toward software features. This step is normally called user requirement specification (URS). The URS step produces a document that specifies requirements for the software, which should be reflected as features or modules for the software (ISM).

The main step of the development of ISM is design. ISM has to fulfill what the users actually need. Since some users are not able to communicate the entirety of their needs, the information they provide may be incomplete, inaccurate and sometimes self-conflicting. The complete information is gathered in the URS stage that specifies users' requirements

---

[4] SDA is a software development process combining elements of both design and prototyping-in-stages, in an effort to combine advantages of top–down and bottom–up concepts.

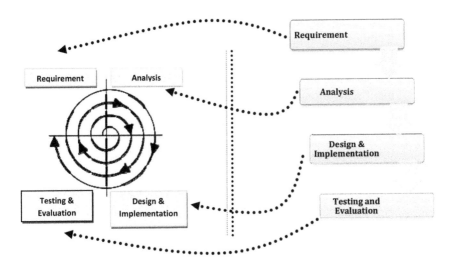

**FIGURE 5.5**   Two essential approaches of software development.

in a specific and unambiguous manner. The well-defined and complete specifications (what the users want) are documented, and then incorporated in the following development stages.

### 5.4.1.1 Functional Requirement

This stage defines modules in the ISM based on functional requirements. Functional requirements define functions of a system or its components. A function is described by a set of inputs, the behavior, and outputs. Functional requirements may be calculations, technical details, the data manipulation, processing and other specific functionalities that define what a system is supposed to accomplish. The details of functional requirements of ISM are shown in Table 5.4.

Verifying user requirements is conducted at the URS stage to find out what features should be incorporated in the software. Several requirements description and claims by users for further ISM development are as follows:

1.  ISM follows the logic flow of ISF so that users might grasp how ISM works to measure and investigate the organization's information

security circumstances. User manual should be provided as it is a technical communication document intended to provide assistance to users.

2. Numerical display module, such as percentage, points, and achievement scale intended to be assembled in ISM.

3. Level-by-level investigation results to be displayed as inputs and advice provided to management for decision making. The investigation results encompass levels of assessment issues, controls, clauses, domains, and top domain.

4. In addition to a numerical style, displays in the form of histograms and graphics should be incorporated for better understanding and ease of analysis, particularly for the board of managers (BoM) and the board of directors (BoD) level. They argue that the medium-upper level management are not concerned with the details and complicated numbers, since it is considered confusing and time-consuming to understand the meaning of those numbers or scores. As a solution, the BoD and BoM stated that the graphics, histograms and images need to be provided.

## 5.4.1.2 Non-Functional Requirement

The non-functional features of the ISM are at least as important to end users. Non-functional features specify criteria that can be used to judge the operation of a system, rather than specific behaviors. This should be contrasted with functional requirements that define specific behavior or functions. In a significant portion, it is the user interface design and style, look and feel of a system that initially attracts users to the software and use what the system has to offer. Lack of non-functional features for many users becomes a reason to abandon of the system. In the non-functional requirement stage, the issues of design, ease of use, layout, and color composition, is often more attractive than the function itself. Several important non-functional requirements of ISM are briefly explained below.

1. *User Friendly.* ISM should be developed using a graphical user interface (GUI) style. The advantages of a graphical user interface or a GUI is not only good for individuals but also good for organizations. Firstly, a GUI allows users to point and click

**TABLE 5.4**   Functional Requirements

| Functional Requirement | Description |
|---|---|
| Conduct RISC evaluation at all layers:<br>§ RISC investigation for controls level<br>§ RISC investigation for clause level<br>§ RISC investigation for domain level<br>§ RISC investigation for top domain level | Users easily perform a review of the infosec circumstances level by level. |
| Strength & weakness descriptions to determine which controls have already been implemented (strengths) and analyzed controls that have not been fully implemented (weaknesses). | Improvement process becomes easier and focused on the controls with weakness statuses |
| Achievement and priority histogram to determine the level of RISC investigations compared with the ideal conditions that should be satisfied to comply with ISO 27001. Gaps that arise between achievement and ideal conditions are correlated with the level of priority for further improvement stage. | By analyzing the gap between the ideal situation and the achievement so far, it will be easier for the organization to make improvements on controls with the biggest gaps as a first priority |
| Comparing the results of RISC investigation on top level domain with the minimum requirements that must be satisfied by an organization to obtain ISO 27001 certification. | Respondent can understand at which stage they are at this time and what scenarios they have to prepare to follow ISO 27001 controls |
| Provide full details on the achievements of this test, which consists of 6-domains, 7-clauses, 21-essential controls, 144-assessment issues, 256-refined questions. Display it in the form of a histogram (figure) and presentation (numerical). | In a comprehensive evaluation of the extent the user can perform all components and controls in ISO 27001 |
| Firewall for controlling the incoming and outgoing network traffic by analyzing the data packets and determining whether they should be allowed through or not, based on a rule set. A firewall establishes a barrier between a trusted, secure internal network and another network (e.g., the Internet) that is not assumed to be secure and trusted. | Users can perform a communication event and data transfer over the network securely with the support of a firewall, especially in DMZ[1] area. |
| Regarding malicious software (malware), which is detected by the firewall, the module "process information" allows users or stakeholders to learn more about the potential suspects. In this module each suspect will be seen in its more detailed properties, such as accessed file names that are running, set size, threads, file type (service/program/port), process ID's and parent process. | Users can use all resources, software and hardware in an organization with the information monitored by the monitoring process |

**TABLE 5.4**  (Continued)

| Functional Requirement | Description |
|---|---|
| Network detection, functioning as workstation monitoring that connects to the entire network, wired or wireless network. Various types of workstation might be a personal computer, tablet, or smart mobile device (smartphone). For instance, the admin would like to know who is currently on a network-connected within the IP: 10.2.1.1 to 10.2.1.255, then the monitor will show the list of names, network-connected machine with IP address. In case there is a machine which is unrecognized as a network member, the admin perpetrates a preventive action to protect related network, data, and information. | Prevent unauthorized users from entering the system network that allows organizations to prevent security breaches. |
| Port Detection, which is an application-specific or process-specific software construct serving as a communications endpoint in a computer's host operating system. A port is associated with an IP address of the host, as well as the type of protocol used for communication. The purpose of ports is to uniquely identify different applications or processes running on a single computer and thereby enable them to share a single physical connection to a packet-switched network like the Internet. Sometimes ports are also used by hackers as a way in to do the demolition on network components, information, and data. Therefore, monitoring of absolute port is required, particularly to prevent the things that are not desired as stated. Securing ports does not mean closing all the ports, because it will instead hinder the flow of traffic and data communication networks. | Make arrangements for the benefit of an open port-specific function of computing and network processes and ensure that they are safe from suspected security breaches |

[1] In computer security, a DMZ or Demilitarized Zone is a physical or logical subnetwork that contains and exposes an organization's external-facing services to a larger and untrusted network, usually the Internet. The purpose of a DMZ is to add an additional layer of security to an organization's local area network (LAN); an external attacker only has direct access to equipment in the DMZ, rather than any other part of the network.

versus a text-based interaction. It is so much easier to point and click than having to use the keyboard to maneuver on the computer screen. The advantage of having a GUI for the company is that it requires less skill. Users do not have to be trained extensively to manage a GUI environment, which saves the company more time than having to train someone for extended periods. The bottom line for GUI systems is productivity, it lessens the amount of time taken to complete a single task. For individuals, the advantages are also great because a graphical user interface takes care of most of the mental processes.

2. *Easy to Use.* Easy to use means that there is low or no barrier for usability and learning for the system. The system will provide multiple interfaces and technology window alternatives, each of which may be of varying levels of sophistication. The user will have the opportunity to choose the option that works best for them.

3. *Personal Authority.* This non-functional feature is an important factor to consider for a successful user interface. The interface of the system should look and feel to the user exactly as the user would want it to look and feel. The user should have a variety of options and styles for the interface, device, and appliance. In the personal-ization issue, user-centricity is one of the founding principles of the system development. It is about involving the user in the use of their identity data, redefining user emphasis, redefining user involve-ment, and creating user models (Patton, 2005). Those issues can be accomplished through flexible ISM design and architecture, deter-mining user preferences, proactively monitoring usage patterns, and assigning intelligent software agents to do the task.

4. *Ease of Access.* A wide range of features should be made available to users on a low-computer specification and normal bandwidth of network.

## 5.4.2 ANALYSIS STAGE

Software development analysis refers to the URS document. The cause-effect analysis matrix is derived from interviews with respondents who are in management and technical positions such as IT manager, IT officer, information security officer, ISO 27001 officer, risk manager, business portfolio and synergy data support officer, ISO 27001 implementation kick off and live test evaluation team, scheduling operational and maintenance manager, system analysis, network and security engineer, programmer, and medical doctor (the details of respective respondents are described in Chapter 3).

The problems, causes-effects, and potential solutions were described at the analysis stage (Table 5.5). From this analysis, a software design and architecture is derived, which is further developed into an appropriate application system that meets user expectations such as being complete, powerful, and user friendly. The development phase will be discussed further in the Subsection 5.4.3.

## 5.4.3 DESIGN AND IMPLEMENTATION STAGE

ISM consists of two major functions: security assessment management (SAM/e-Assessment) and Security monitoring management (SMM/e-Monitoring). SAM is to measure ISO 27001 parameters, i.e. RISC investigation, based on the proposed framework (Figure 5.6). It contains 6 domains, 7 clauses, 21 essential controls and 144 assessment issues. ISM has created opportunities to overcome the obstacles that may be faced during ISO 27001 compliance stages.

ISM is equipped with a login system to track user activity and organizations' RISC investigation and allows recognition of assessment patterns. This assessment pattern may reflect an organization's readiness that leads to enhancement of the interactions between the organization and its consultant to define the scope of controls for ISO 27001 compliance. ISM consists of 80 object modules, 80 user interfaces, 162–420 lines programming code per-page, 16,000 lines of programming code, and SQL[5] as the database-bridge to GUI. Some additional features were developed as support features for user explanation: (1) acknowledgement, (2) team, (3) help, (4) global survey, and (5) go online (Figures 5.7–5.10).

ISM helps users to unveil ISO 27001 terms and controls that are very difficult to understand (Alfantookh, 2009; Susanto et al., 2012a) and assists users in understanding RISC investigation, reduce dependency on consultants, which may lead to increased efficiency and cost reduction associated with compliance processes (Kosutic, 2010, 2012).

---

[5] Structured Query Language is a special-purpose programming language designed for managing data held in a relational database management system (RDBMS).

**TABLE 5.5** Analysis and Solution for ISM

| Problem | Causes and Effects | Solution |
|---|---|---|
| The lack of a formal method to measure organizational readiness level for ISO 27001 compliance. | Organizations are dependent upon a consultant to understand terms, concepts, SoA preparation and stages for compliance with ISO 27001. | Propose an applied framework and transform it into a software application to conduct RISC investigation. |
| The lack of software applications that can help organizations understand their information security positions and circumstances with respect to ISO 27001 | Organizations can only use the traditional method, which is the checklist implementation method (ICM) to pursue the parameters of the standard. The purposes of ICM are: (1) to gauge the level of compliance with ISO 27001 requirements by holding, department, and division. (2) Facilitate the provision of information for ISO 27001 implementation. (3) Serve as training materials for understanding the ISO 27001 requirements. | There is a need to develop software that can practically help organizations measure their readiness for the implementation of ISO 27001 |
| Need to develop software that is powerful, easy to use, and graphics based. | The advantages of a graphical user interface or GUI are not only for individuals but also for organizations. A GUI allows user to point and click versus text-based interactions. Users do not have to be trained extensively to manage a GUI environment, which saves the company more time for training. The bottom line for GUI systems is the productivity; it lessens the amount of time take to complete a single task. For individuals, the advantages are also great because the graphical user interface takes care of most of the mental processes[2]. | Need to develop GUI-based software (ISM) |

**TABLE 5.5** (Continued).

| Problem | Causes and Effects | Solution |
|---|---|---|
| Users find it difficult to investigate their information security circumstances in detail level-by-level, such as assessment issues, controls, clause, and domain. | There is no methodology that can measure in detail as layer-by-layer for organizations to better focus through strengths-weaknesses analysis. | Develop a module that can automatically measure the level of organizational readiness in detail layer-by-layer, and strengths-weakness analysis. |
| Organizations find it difficult to focus on their weak parts of the achievement. | Currently, the organization uses ICM tools, the consequence is that they need a considerable amount of time to review all controls to find their weak points. Therefore, there is a need for software that can systematically control the classification based on topic and function, in order to easily perform an analysis of the organization's overall controls that must be evaluated and implemented. | Develop features that can directly show the weak points of an organization towards ISO 27001 implementation, so that organizations can quickly perform improvement on that point. |
| Respondets need monitoring system to conduct and check their security in real-time against suspected information security breaches. | No integrated systems were developed as measurement system (RISC investigation) and information security breach suspect monitoring system in real-time. | Develop a monitoring system that integrates with the RISC investigation so that respondents could also conduct RISC investigation and security monitoring at the same time. |

[2] Mental process or mental function are terms often used interchangeably for all the things that individuals do with their minds. These include perception, memory, creativity, imagination, and ideas.

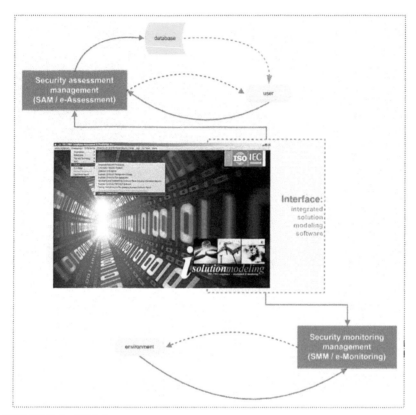

**FIGURE 5.6**    ISM diagram.

### 5.4.3.1   The System Development

One of the most important phases in software development is the programming phase which involves various aspects of programming languages, database software, and strategy implementation for procedures and functions to be effective, reusable, and inheritable in the other parts of the program. The ISF implementation proposes to provide a foundation to build an integrated solution that can incorporate organization enhancement electronically and make it a paperless process. The major benefits of the software as it is integrated solution for analyzing risk, reduce compliance cost, and shorten time for compliance preparation.

**FIGURE 5.7**   Main page.

## 5.4.3.2   ISM Sub Systems

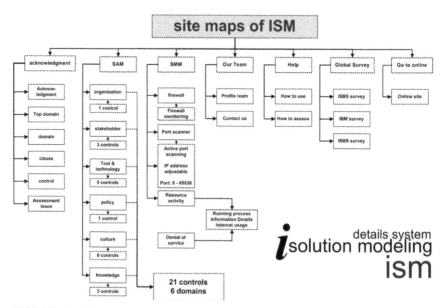

**FIGURE 5.8**   Site map.

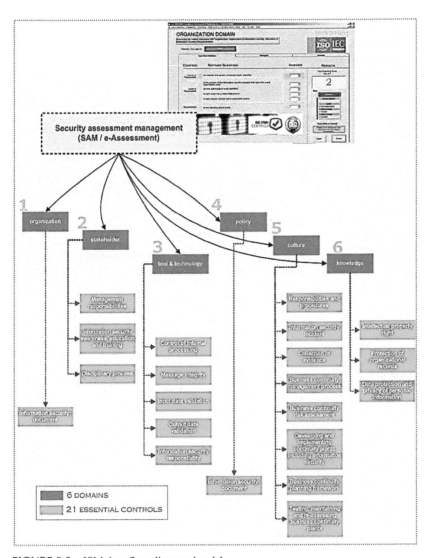

**FIGURE 5.9** ISM data flow diagram level 1.

The details of security assessment management (SAM/e-Assessment) and security monitoring management (SMM/e-Monitoring) are explained in the following subsections.

**FIGURE 5.10**   Sign up.

### 5.4.3.2.1   Main Page

The main page of ISM is designed to be user friendly with a clear navigation design to allow users to find and access information effectively. The main menu consists of acknowledgement, e-assessment, e-monitoring, help, global survey on information security trends, login/logout (Figures 5.8 and 5.9). Acknowledgement consists of information about the ISM such as software functions, the mathematical notations formula and brief information on key benefits of ISO 27001 implementation.

E-assessment (SAM) contains a submenu with the items Organization, Stakeholder, Tool and technology, Culture, Knowledge, Policy, and Top domain. These submenus function as direct links to the electronic assessment form dealing with each domain. A pull-down menu also exists in each submenu, which is the controls menu. The controls menu consists of 21 essential controls of ISO 27001, and the menu structure follows ISO 27001 components as mapped in ISF.

E-monitoring (SMM) is a representation of real-time network monitoring for suspected information security breaches. Under the e-monitoring menu, there are several direct links to the monitoring functions: firewall detection, port scanning and resource activity denial of service. Firewall detection consists of firewall monitoring and detection in the submenu. Port scanning has Active port detection, Open port status, and IP address adjustable in its submenu. In the Resource activity denial of service, the submenu contains running processes, information details and internet usage. The details of each submenu item and its functions will be further explained in Subsection 5.4.3.2.2.

The help menu explains how to use ISM, like a user manual of the software. The "Global survey on information security trends" menu item is based on ISBS surveys of the 2008, 2010, 2012 trends and tendencies regarding information security breaches. Logout functions as a single "lock" of the system without requiring user confirmation to logout.

### 5.4.3.2.2   Security Assessment Management (SAM/e-Assessment)

Security assessment management (SAM/e-Assessment) provides electronic assessment or RISC investigation. This electronic assessment consists of refined questions, assessment issues, controls, clauses, domains, and the top domain (Figure 5.7). SAM assesses organizations on their information security management and technical strategy.

The assessment produces SoA[6] (previously discussed in Chapter 2, *Literature Review*) documentation that identifies the controls chosen and why it is appropriate. The purpose of SoA is to list all controls and to define those which are applicable and those which are not, and the reasons for such a decision, the objectives to be achieved with the controls and a description of implementation. SoA is the central document that defines how an organization has or will implement information security controls. It is to define which of the suggested 133 controls, including 21 essential controls of security measures, will apply, and for those that are applicable, how they will be implemented. In the current practice, if an organization goes for ISO 27001 certification, an auditor takes the SoA

---

[6] SoA is the central document that defines how an organization will implement (or has implemented) information security measures. SoA is the main link between the risk assessment and treatment and the implementation of information security to define which of the suggested 133 controls (included 21 essential controls of security measures) will apply, and how to implement it. Actually, if an organization goes for ISO 27001 certification, the certification auditor will take the SoA and check around the organization/company on whether it has implemented the controls in the way described in it.

report and check whether the organization has implemented controls in the way described in the SoA.

### 5.4.3.2.2.1   User Sign Up

Sign-up is the basic stage for the user to become a member of the system. One of the benefit become user-member is the ability to monitor the investigation pattern during RISC investigation and network monitoring. In order to complete the sign-up process, users need to confirm their username and password. THE information needed for registration are only username, password, and re-confirm password, for the person in-charge as the organization's representative (Figure 5.10).

### 5.4.3.2.2.2   Login

User(s) are required to key in their user name and password, and then sign in. When users attempt to sign in with a user name and the password does not match, the system returns an error. ISM keeps users signed in until they explicitly sign out. The Log in menu contains miscellaneous commands that mainly provide command to tune user account settings and display information about the ISM interface (Figure 5.11).

### 5.4.3.2.2.3   Knowledge and Information

The Knowledge menu comprises a range of knowledge resources focused on information security and related topics. Users could access knowledge and information to seek any topics related to the information security standard, ISO 27001, information security breaches (Figure 5.12), key benefits for ISO 27001 implementation (Figure 5.13), and self-assessment stages (Figure 5.14). The knowledge and information menu will empower and enhance users to build their information security literacy. The pages provide comprehensive information security advice and knowledge to help users expand their literacy scope.

Knowledge and information is a learning center for users to learn about further processes of compliance. The source of knowledge is from

**FIGURE 5.11**    Login.

literature and experts which have been verified to ensure reliability. The interface as shown in Figures 5.12–5.14, and these pages also function as a knowledge and information warehouse. In this feature the admin has the full authority to create, add and update information.

### 5.4.3.2.2.4    Assessment Module

The Assessment module represents the RISC investigation. On this electronic form, the user is prompted to fill the achievements for each of the assessment issues. The level of assessment scoring is set out in range of 0 to 4: 0 = not implementing, 1 = below average, 2 = average, 3 = above average, and 4 = excellent.

As a measurement example, several steps of parameter assessment are described as follows (Figure 5.15):

1.  Domain: "Organization".
2.  Controls: "Organization of information security: Allocation of Information Security Responsibilities".

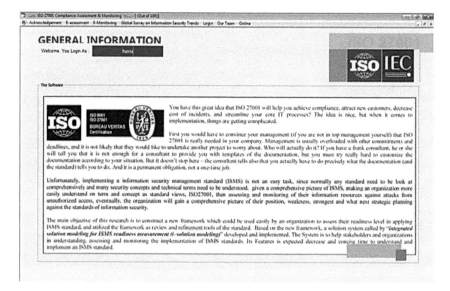

**FIGURE 5.12**   Feature: highlights of ISO 27001

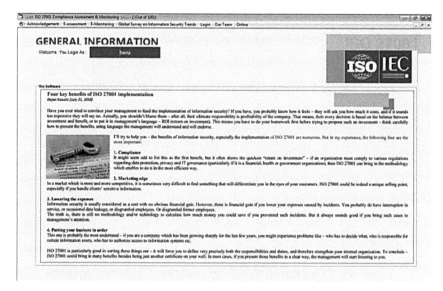

**FIGURE 5.13**   Feature: key benefits of ISO 27001 implementation.

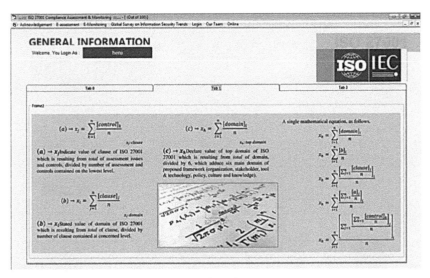

**FIGURE 5.14**   Feature: the investigation formula.

3. Assessment issue: "Are assets and security process cleary identified?"
4. Then stakeholders should analogize the ongoing situation, implementation and scenario in the organization, and benchmark it to the security standard's level of assessment.

## *5.4.3.2.2.5   Histogram Module*

This module provides the details of the organization's achievement and priority in a histogram format. The strength and weakness points on an organization's current achievement will be shown. As indicated in the system, *"Achievement"* declares the performance of an organization as a final result of the measurement validated by the framework. *"Priority"* indicates the gap between the ideal score and the achievement value. "Priority" and "achievement" shows an inverse relationship. If achievement of a domain is high, then the domain has a low priority for further improvement, and conversely, if achievement of a domain is low, then the priority for the domain is high (Figure 5.16).

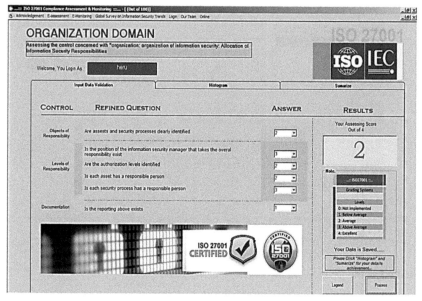

**FIGURE 5.15**    Assessment form.

## 5.4.3.2.2.6    Domain Result Module

The Submenu item "Domain result" has a feature that provides the user analysis of their achievement, as shown in Figure 5.17. Some of the assessment criteria are displayed and in addition advice is prepared, and it gives advice based on the previous assessment. There are four main features of this module as follows:

- Final result out of 4 scale.
- Final result out of 100%.
- Final achievement of assessment result (not implementing, below average, average, above average, excellent).
- Advice from the software regarding their final achievement, which points to their strongest areas and also their weakest areas.

The user conducts RISC investigation for his company by filling in an electronic form displayed in the application. The results of this investigation can then be referred to as an overall evaluation of the organization's information security circumstances, which would be compared to the ideals in the ISO 27001 standard.

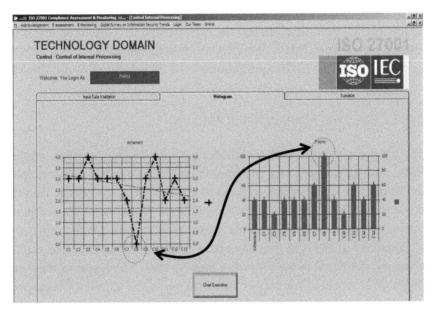

**FIGURE 5.16** Final result view on histogram style.

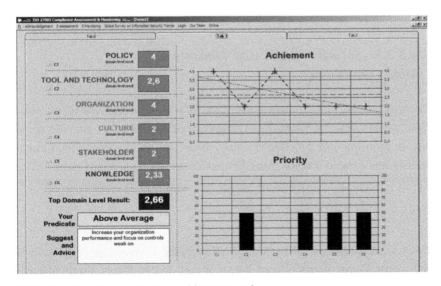

**FIGURE 5.17** Final result view on histogram style.

In some cases, the organization needs more than one experiment to conduct RISC investigation, and the time needed for each experiment is about 30 to 60 min. This is definitely much better than the other common approaches, which require approximately 12–24 months (Kosutic, 2013).

### 5.4.3.2.2.7   Top Domain Result Module

The top domain result module provides a comprehensive overview on the RISC investigation result for an organization. We have provided an example to illustrate the measurement process to reveal how ISM works in the top domain level. Each question of the refined simple elements and their respective values is given in the example. Figure 5.18 summarizes the results of all domains together with their associated controls based on ISF. The results given are illustrated in the following figures. Table 5.6 represents the condition of 21 essential controls of standards and also Figure 5.19 states the overall condition of 21 essential controls in a histogram style. The overall score of all domains is shown in the Table 5.6, to be "2.66 points". The domain of the "policy" scored highest at "4", and the domain of the "knowledge" scored lowest at "2". Ideal and priority figures are given to illustrate the strongest and weakest in the application of each control (Figure 5.20).

### 5.4.3.2.3   Security Monitoring Management (SMM/e-Monitoring)

Security monitoring management (SMM/e-Monitoring) monitors real-time system activity, firewall and network management to provide security monitoring and potential security breaches. ISM tools collect event data in real time to enable immediate analysis and response (Figure 5.21).

### 5.4.3.2.3.1   Firewall Management

Firewall management (FM) is a module-based network security system that controls the incoming and outgoing network traffic by analyzing the data packets based on a set of rules and subsequently determine whether

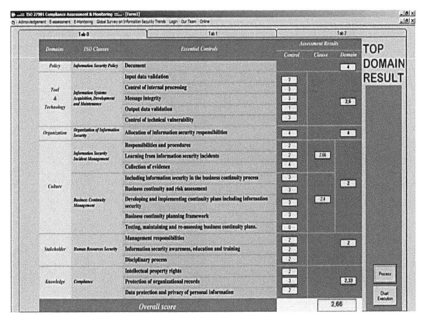

**FIGURE 5.18**   Final result view in summary style.

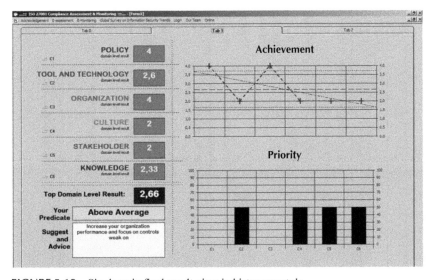

**FIGURE 5.19**   Six domain final result view in histogram style.

**TABLE 5.6** RISC Measurement

| | | | Summarize | | | |
|---|---|---|---|---|---|---|
| | | Assessed Items | | Assessment Result | | |
| **Domain** | **Clauses** | **Controls** | **Control** | **Clause** | **Domain** |
| 1 Organization | Organization of information security | Allocation of information security responsibilities | 4 | 4 | 4 |
| 2 Stakeholder | Human resources security | Management responsibilities | 2 | 2 | 2 |
| | | Information security awareness, education and training | 2 | | |
| | | Disciplinary process | 2 | | |
| 3 Tools & Technology | Information system acquisition, development, and maintenance | Input data validation | 3 | 2.6 | 2.6 |
| | | Control of internal processing | 3 | | |
| | | Message integrity | 3 | | |
| | | Output data validation | 1 | | |
| | | Control of technical vulnerability | 3 | | |
| 4 Policy | Information security policy | Document | 4 | 4 | 4 |
| 5 Culture | Information security incident management | Responsibilities and procedures | 2 | 2.66 | 2 |
| | | Learning from information security incident | 2 | | |

| | | Collection of evidence | | |
|---|---|---|---|---|
| | | 4 | | |
| Business continuity management | Including information security in the business continuity process | 3 | 2.4 | |
| | Business continuity and risk assessment | 3 | | |
| | Developing and implementing continuity plans including information security | 3 | | |
| | Business continuity planning framework | 3 | | |
| | Testing, maintaining and re-assessing business continuity plans | 0 | | |
| 6 Knowledge | Compliance | Intellectual property rights | 2 | 2.33 | 2.33 |
| | | Protection of organization record | 3 | | |
| | | Data protection and privacy of personal information | 2 | | |
| Top Domain Overall Score | | | | 2.66 |

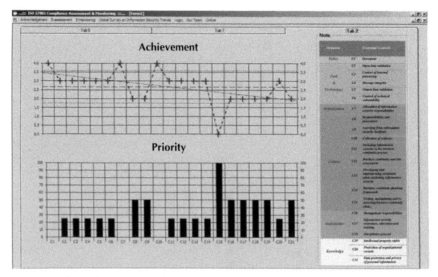

**FIGURE 5.20**   The 21 essential controls final result view in histogram style.

they are allowed to pass through or not. A firewall establishes a barrier between a trusted secure internal network and another network (e.g., the Internet), which is assumed insecure and not trusted.

FM is frequently used to prevent unauthorized Internet users from accessing private networks connected to the Internet, especially *intranets*. All messages entering or leaving the network pass through the firewall, which examines each message and blocks those that do not meet the specified security criteria (Figure 5.22).

In ISM, firewall management represents an essential control in the standard, which functions as a monitoring tool for traffic activity programs concerned with a particular network. It is also as a tool to maintain network activity, to follow procedures and guidance of information security scenarios. Moreover, FM manages files that are executed by a computer machine either coming out or coming in to the network and the respective workstation. If it is considered suspicious, such as iteration and traffic process to access a file or application in a network or machine for instance that exceeds 100 Mb.secon$^{-1}$ or more than 1000 access.secon$^{-1}$, this would indicate a traffic anomaly, then the system

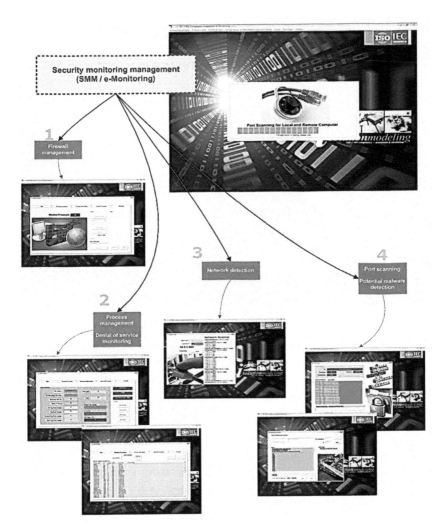

**FIGURE 5.21** Security monitoring management.

gives an indication signal showing "blocked processes" as a preventive scenario to assure security of the computer and network.

The FM in ISM uses kernel (32) facilities to analyze memory management, to determine whether the memory usage is still within reasonable limits to run a number of software and data traffic into and out of the system. Kernel (32) is the central module that contains the core processes or

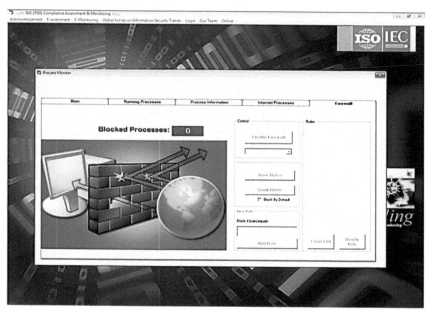

**FIGURE 5.22**   Firewall management.

heart of the operating system. At boot, the kernel (32) loads into memory, regulating operations as the user runs various tasks and programs. Kernel (32) is the 32-bit dynamic link library (.dll) found in the operating system (OS) kernel. It handles memory management, input-output operations, and interrupts (Table 5.8).

A "dll" file can be accessed by more than one program at a time, so the kernel32.dll serves not only the operating system, but can also accommodate installed third-party programs. Additionally, the kernel (32) file regulates memory management, input and output streams, necessary task management and disk management. When kernel32.dll loads into memory, it protects the address field or "page" it occupies to keep other programs from stealing memory. Sometimes it happens that software will attempt to access this memory page, triggering an "invalid page fault error". If a particular program continually produces a page fault error message, it probably indicates a security hole.

The early system indicator is absolutely needed to ensure the network traffic is under control, to prevent viruses, worms, or hackers getting into the system. As it is described in ISO 27001 in the domain "tools

& technology", clause information systems acquisition, development and maintenance, particularly in the control "information systems acquisition, development and maintenance: control of technical vulnerabilities", essential control number "12.6.1" in the standard (Table 5.7), that a system must have rules, scenarios and techniques as deterrence against possible vulnerabilities in an organization's computer network.

### 5.4.3.2.3.2   Process Information

To further analyze malicious software (malware) detected by the firewall, as mentioned previously, it is needed to know more details about the network activity to find out suspected security breaches such as malware, botnet, and denial of service. The module "process information" allows users to learn more about the potential suspect processes. In this module each suspect will be seen in more detail, such as: file name accessed, set size, threads, file type (service/program/port), process ID and parent process (Figures 5.23 and 5.24).

**TABLE 5.7**   Essential Control Number 12.6.1: Control of Technical Vulnerabilities

*Timely information about technical vulnerabilities of information systems being used should be obtained, the organisation's exposure to such vulnerabilities evaluated and appropriate measures taken to address the associated risk.*

| Assessment Issue | Refined Question |
|---|---|
| Inventory of technical assets | Do the technical specifications of systems and their components exist? |
| Vulnerability | Are the vulnerabilities of technical assets identified? |
| | Are the risks associated with vulnerabilities identified? |
| Protection | Are protection measures that respond to risks well-identified? |
| | Are the protection tools evaluated before use? |
| | Does awareness on potential vulnerabilities among the right people exist? |
| Practice | Does the monitoring to manage problems exist? |
| | Does logging of events exist? |
| Accountability | Do the defined responsibilities exist? |
| Documentation | Does reporting of the above exist? |

**TABLE 5.8** Firewall Management

```
Public Declare Function CreateToolhelp32Snapshot Lib "kernel32"
Public Declare Function Process32First Lib "kernel32"
Public Declare Sub CloseHandle Lib "kernel32"
Public Sub enumProc()
Dim procType As String
procType = ""
servProc = 0
uknProc = 0
sysProc = 0
If monitor <> 1 Or firewallStatus <> 1 Then
If noClear = 0 Then
frmProc.lstvwProc.ListItems.Clear
End If
Else
If tempClear = 1 Then
frmProc.lstvwProc.ListItems.Clear
tempClear = 0
End If
End If
Dim hSnapShot As Long, uProcess As PROCESSENTRY32
hSnapShot = CreateToolhelp32Snapshot(TH32CS_SNAPALL, 0&)
uProcess.dwSize = Len(uProcess)
r = Process32First(hSnapShot, uProcess)
r = Process32Next(hSnapShot, uProcess)
Do While r
ProcessName = Left$(uProcess.szExeFile, IIf(InStr(1, uProcess.szExeFile, Chr$(0)) >
0, InStr(1, uProcess.szExeFile, Chr$(0)) - 1, 0))
If UCase(ProcessName) = UCase("services.exe") Then
servPID = uProcess.th32ProcessID
ElseIfUCase(ProcessName) = UCase("explorer.exe") Then
expPID = uProcess.th32ProcessID
ElseIfUCase(ProcessName) = UCase("system") Then
sysPID(1) = uProcess.th32ProcessID
ElseIfUCase(ProcessName) = UCase("smss.exe") Then
sysPID(2) = uProcess.th32ProcessID
```

**TABLE 5.8**   (Continued)

---

ElseIfUCase(ProcessName) = UCase("winlogon.exe") Then

sysPID(3) = uProcess.th32ProcessID

ElseIfUCase(ProcessName) = UCase("csrss.exe") Then

sysPID(4) = uProcess.th32ProcessID

ElseIfUCase(ProcessName) = UCase("lsass.exe") Then

sysPID(5) = uProcess.th32ProcessID

End If

If popView = 1 Then

If uProcess.th32ParentProcessID = servPID Then

servProc = servProc + 1

procType = "Service"

ElseIf uProcess.th32ParentProcessID = expPID Then

uknProc = uknProc + 1

procType = "Unknown"

ElseIf uProcess.th32ParentProcessID = sysPID(1) Or uProcess.th32ParentProcessID = sysPID(2) Or uProcess.th32ParentProcessID = sysPID(3) Or uProcess.th32ParentProcessID = sysPID(4) Or uProcess.th32ParentProcessID = sysPID(5) Then

sysProc = sysProc + 1

procType = "System"

End If

Call popLstvw(ProcessName, uProcess, getPriority(uProcess.th32ProcessID), getProcMem(uProcess.th32ProcessID), getCTime(uProcess.th32ProcessID), procType)

End If

---

By looking at the file's properties and details, it becomes easier for users to further analyze the suspected file, by analyzing the traffic processes and file repetition accesses by malicious software. This module also comes with some additional features, such as: set priority process and kill process. "Kill process" aims to stop a running process if it is suspected to endanger the system, called denial of service (DoS).

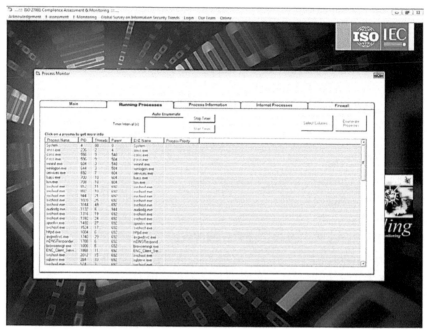

**FIGURE 5.23**    Running processes.

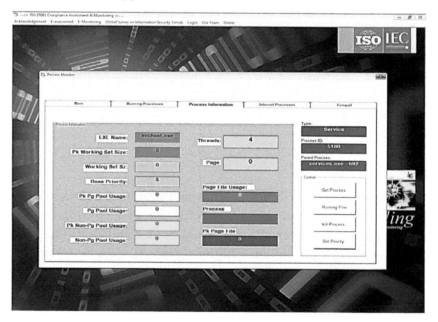

**FIGURE 5.24**    Process information.

A DoS attack is an attempt to make a machine or network resource unavailable to its intended users. Although the means to carry out, motives for, and targets of a DoS attack may vary, it generally consists of efforts to temporarily or indefinitely interrupt or suspend services of a host connected to the network. One common method of attack involves saturating the target machine with external communications requests. In general terms, DoS attacks are implemented by either forcing the targeted computer(s) to reset, or consuming its resources so that it can no longer provide its intended service, or obstructing the communication media between the intended users and the victims can no longer communicate adequately.

This module is an implementation of the domain "tools & technology" clause "Information Systems Acquisition, Development and Maintenance" and the control "data validation input" and "output validation data" in essential control number 12.2.1 (Table 5.9) and 12.2.2 (Table 5.10).

### 5.4.3.2.3.3    Network Detection

Network detection functions as workstation monitoring that connects to the entire network, be it wired or wireless. Various types of hardware might be connected to the network such as a personal computers, tablets, or smart mobile devices (smartphones). For instance, if a system administrator would like to know who is currently on the network within the IP range of 10.2.1.1 to 10.2.1.255, the monitor will show a list of names of network-connected machines with their IP addresses. In case there is a machine which is unrecognizable as a network member, then the system administrator perpetrates preventive action to protect related network, data, and information (Figure 5.25).

If an intruder penetrates into an organization's network, it may lead to information scrounging and will adversely influence the customers' trust, as it indicates that the organization is vulnerable. In accordance with the control and assessment issues in the standard, this module represent the "culture" domain, which describes the organization's culture strategy towards information security challenges (Table 5.11), and assessing the control in *culture: business continuity management: including information security in the business continuity*

**TABLE 5.9** Correct Processing in Applications: Input Data Validation

| *Data input to applications should be validated to ensure that this data is correct and appropriate* | |
| --- | --- |
| **Assessment Issue** | **Refined Question** |
| Existence | Do plausibility checks exist to test the output data reasonability? |
| Validation | Does examination for the input business transaction, standing data and parameter tables exist? |
| | Does automatic examination exist? |
| | Are periodic reviews and inspection available? |
| | Do response procedures to validation exist? |
| Management | Does logging of events exist? |
| Accountability | Are the responsibilities defined? |
| Documentation | Does reporting of the above exist? |

**TABLE 5.10** Information Systems Acquisition, Development and Maintenance: Output Data Validation

| *Data output from an application should be validated to ensure that the processing of stored information is correct and appropriate to the circumstances* | |
| --- | --- |
| **Assessment Issue** | **Refined Question** |
| Existence | Do plausibility checks exist to test the output data reasonability? |
| Validation | Is the provided information for a reader or subsequent processing system sufficient to determine the accuracy, completeness, precision and classification of the information? |
| | Do periodic inspections exist? |
| | Do response procedures to validation tests exist? |
| Practice | Does the checking that programs are run in order exist? |
| | Does the checking that programs are run at the correct time exist? |
| Accountability | Are the responsibilities defined? |
| Documentation | Does reporting of the above exist? |

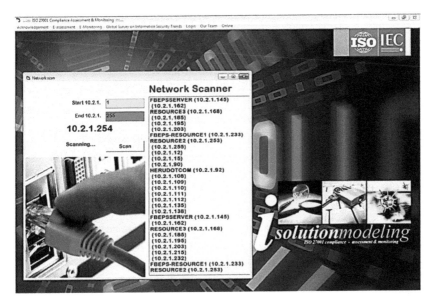

**FIGURE 5.25**    Process information.

*management process and business continuity management: business continu-ity and risk assessment* (Table 5.12).

### 5.4.3.2.3.4   Port Scanning for Local and Remote Network

A port is an application-specific or process-specific software construct serv-ing as a communications endpoint in a computer's host operating system. A port is associated with an IP address of the host, as well as the type of protocol used for communication. The purpose of ports is to uniquely identify different applications or processes running on a single computer and thereby enable them to share a single physical connection to a packet-switched network like the Internet. Ports are also used by hackers to destruct network components, information and data. Therefore, monitoring of ports is required, particularly to prevent the things that are not desired as stated. Securing the port does not mean closing all the ports, because it will instead hinder the flow of traffic and data communication in the network (Figures 5.26–5.28).

**TABLE 5.11**   Including Information Security in the Business Continuity Management Process

| *A managed process should be developed and maintained for business continuity throughout the organization that addresses the information security requirements needed for the organization's business continuity* | |
|---|---|
| **Assessment Issue** | **Refined Question** |
| **Identification of critical business processes** | Are the critical business processes identified? |
| | Are the critical business processes prioritized? |
| | Are the assets associated with critical business processes identified? |
| **Identification of risk** | Are the risks identified? |
| | Are the probabilities of risks identified? |
| | Are the impacts of risks on business identified? |
| | Are the risks (incidents) classified according to their level of seriousness? |
| **Protection considerations** | Are the information security requirements identified? |
| | Is the protection of personnel identified? |
| | Is the protection of processing facilities identified? |
| | Is the protection of organizational property identified? |
| | Is purchasing insurance considered? |
| **Procedures** | Do procedures considering the above exist? |
| | Are the procedures updated regularly? |
| | Are the procedures incorporated in the business process? |
| **Responsibilities** | Are the business continuity responsibilities assigned at the appropriate levels? |
| **Documentation** | Does reporting of the above exist? |

There is an abundance of available ports, from port "1" to port number "65535", which is a permutation of 216 ($2^{216}$) (Table 5.13). Port 21 is the port that is commonly used for broadband Internet and data communications traffic, which is very common and always open. Moreover, this module provides a tool to detect which ports are open and used for such service. If it is deemed dangerous and threatens the security of information, the user can close the port.

The protocols that primarily use ports are the Transport Layer protocols, such as the Transmission Control Protocol (TCP) and the User

**TABLE 5.12** Business Continuity Management: Business Continuity and Risk Assessment

| Events that can cause interruptions to business processes should be identified, along with the probability and impact of such interruptions and their consequences for information security | |
| --- | --- |
| **Assessment Issue** | **Refined Question** |
| **The organization** | Are the organization's objectives considered? |
| | Are the organization's priorities and criteria considered? |
| | Are allowable outage times identified? |
| | Are the critical resources identified? |
| **Interruption events** | Are the events (i.e., equipment failure, human errors and theft) that cause interruption identified? |
| | Are the probabilities of events identified? |
| | Are the interruption impacts of events identified? |
| | Are the recovery periods identified? |
| | Are the priority risks identified? |
| **Business continuity plan** | Are the information security requirements identified? |
| | Is the protection of personnel established? |
| | Is the protection of processing facilities established? |
| **Documentation** | Does reporting of the above exist? |

Datagram Protocol (UDP) of the Internet protocol suite. A port is identified for each address and protocol by a 16-bit number, commonly known as the port number. The port number, added to a computer's internet protocol (IP) address, completes the destination address for a communications session. Data packets are routed across the network to a specific destination IP address, and then, upon reaching the destination computer, are further routed to the specific process bound to the destination port number.

Note that it is the combination of IP address and port number together that must be globally unique. Thus, different IP addresses or protocols may use the same port number for communication, e.g., on a given host or interface UDP and TCP may use the same port number, or on a host with two interfaces, both addresses may be associated with a port having the same number.

Applications implementing common services often use specifically reserved, well-known port numbers for receiving service requests from

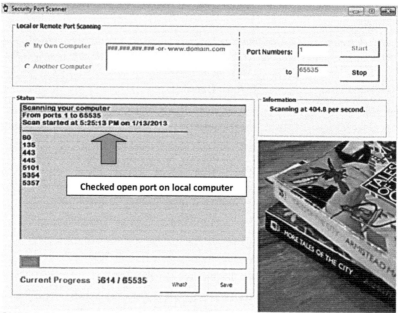

FIGURE 5.26    Port scanning for local computer.

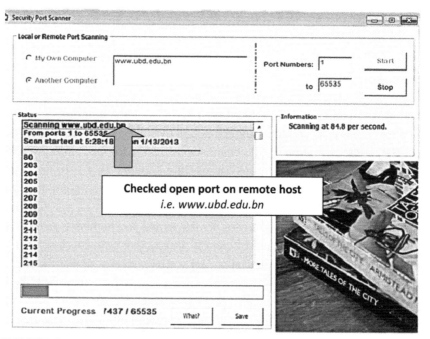

FIGURE 5.27    Port scanning for remote computer.

**FIGURE 5.28**   Port scanning for remote computer.

**TABLE 5.13**   Port Number and Function

| Number | Function |
| --- | --- |
| 20 & 21 | File Transfer Protocol (FTP) |
| 22 | Secure Shell (SSH) |
| 23 | Telnet remote login service |
| 25 | Simple Mail Transfer Protocol (SMTP) |
| 53 | Domain Name System (DNS) service |
| 80 | Hypertext Transfer Protocol (HTTP) used in the World Wide Web |
| 110 | Post Office Protocol (POP3) |
| 119 | Network News Transfer Protocol (NNTP) |
| 143 | Internet Message Access Protocol (IMAP) |
| 161 | Simple Network Management Protocol (SNMP) |
| 443 | HTTP Secure (HTTPS) |

client hosts. This process is known as listening and involves the receipt of a request on the well-known port and establishing a one-to-one server-client connection, using the same local port number; other clients may continue to connect to the listening port. The core network services, such as the World Wide Web, typically use small port numbers less than 1024. In many operating systems, special privileges are required for applications to bind to these ports because these are often deemed critical to the operation of IP networks. Conversely, the client end of a connection typically uses a high port number allocated for short-term use, therefore called an ephemeral port.

The port numbers are encoded in the transport protocol packet header, and they can be readily interpreted not only by the sending and receiving computers, but also by other components of the networking infrastructure. In particular, firewalls are commonly configured to differentiate between packets based on their source or destination port numbers. Port forwarding is an example application of this.

The practice of attempting to connect to a range of ports in sequence on a single computer is commonly known as port scanning. This is usually associated either with malicious cracking attempts or with network administrators looking for possible vulnerabilities to help prevent such attacks. Port connection attempts are frequently monitored and logged by computers. The technique of port knocking uses a series of port connections (knocks) from a client computer to enable a server connection. In accordance with the essential control and assessment issues in the standard, this module represents the knowledge and culture domain (Tables 5.14–5.17).

## 5.5   RELEASE VERSION

Release management is employed to manage the release, distribution, testing and evaluation of software versions. Release categories include major software releases, normally containing large amounts of new functionality, a major upgrade or release usually superseding all minor upgrades, releases and emergency fixes. Minor software releases and hardware upgrades normally contains small enhancements and fixes, some of which may have

**TABLE 5.14**   Compliance: Data Protection and Privacy of Personal Information

*Timely information about technical vulnerabilities of information systems being used should be obtained, the organization's exposure to such vulnerabilities evaluated and appropriate measures taken to address the associated risk*

| Assessment Issue | Refined Question |
|---|---|
| Inventory of technical assets | Do the technical specifications of systems and their components exist? |
| Vulnerability | Are the vulnerabilities of technical assets identified? |
| | Are the risks associated with vulnerabilities identified? |
| Protection | Are protection measures that respond to risks well-identified? |
| | Are the protection tools evaluated before use? |
| | Does awareness on potential vulnerabilities among the right people exist? |
| Practice | Does the monitoring to manage problems exist? |
| | Does logging of events exist? |
| Accountability | Are responsibilities defined? |
| Documentation | Does reporting of the above exist? |

**TABLE 5.15**   Compliance: Data Protection and Privacy of Personal Information

*Data protection and privacy should be ensured as required in relevant legislation, regulations, and if applicable, contractual clauses*

| Assessment Issue | Refined Question |
|---|---|
| Legislations on: collection, processing and transmission of personal info | Do the country level legislations exist: information privacy policy available? |
| | Does an organization level privacy policy exist? |
| | Do the technical protection measures exist? |
| | Do the protection procedures exist? |
| | Are the legislations observed? |
| Responsibilities and practices | Is the data protection officer appointed? |
| | Are the managers' responsibilities and guidelines identified and applied? |
| | Are the user's responsibilities and guidelines identified and applied? |
| | Are the service providers' responsibilities and guidelines identified and applied? |
| Documentation | Does reporting of the above exist? |

**TABLE 5.16**   Information Security Incident Management: Responsibilities and Procedures

| *Management responsibilities and procedures should be established to ensure a quick, effective, and orderly response to information security incidents* | |
|---|---|
| **Assessment Issue** | **Refined Question** |
| Responsibility | Are the procedures concerned with information security incidents approved by management? |
| | Are the commitment levels to the procedures concerned with security incidents identified? |
| Types of incidents | Are management procedures established for each key incident? |
| | Are all key incidents identified? |
| Producers | Do the Incident procedures consider business integrity? |

**TABLE 5.17**   Business Continuity Management Process

| *A managed process should be developed and maintained for business continuity throughout the organization that addresses the information security requirements needed for the organization's business continuity* | |
|---|---|
| **Assessment Issue** | **Refined Question** |
| Identification of critical business processes | Are the critical business processes identified? |
| | Are the critical business processes prioritized? |
| | Are the assets associated with critical business processes identified? |
| Identification of risk | Are the risks identified? |
| | Are the probabilities of risks identified? |
| | Are the impacts of risks on business identified? |
| | Are the risks (incidents) classified according to their level of seriousness? |
| Protection considerations | Are the information security requirements identified? |
| | Is the protection of personnel identified? |
| | Is the protection of processing facilities identified? |
| | Is the protection of organizational property identified? |
| | Is purchasing insurance considered? |
| Procedures | Are there procedures considering the above? |
| | Are the procedures updated regularly? |
| | Are the procedures incorporated in the business process? |
| Responsibilities | Are the business continuity responsibilities assigned at the appropriate levels? |
| Documentation | Does reporting of the above exist? |

already been issued as emergency fixes. A minor upgrade or release usually supersedes emergency fixes. Emergency software and hardware fixes normally contain the corrections to a small number of known problems. ISM has gone through seven versions, released in a sequence of improvement based on stakeholder feedback during the user requirement stage within software development, and also during the testing stage. Those versions are as follows:

1. Version 1: This version consisted of 1 domain: Tool & Technology → April 2011.
2. Version 1.1: This version consisted of 1 domain: Tool & Technology, with adjustments and refinement formula → May 2011.
3. Version 2.0: This version consisted of 6 domains: Tool & Technology, Organization, Policy, Knowledge, Culture, and Stakeholder → August 2011.
4. Version 3.0: This version consisted of 6 domains: Tool & Technology, Organization, Policy, Knowledge, Culture, and Stakeholder. In this version, SMM, the monitoring system, was introduced → October 2011.
5. Version 4.0: This version consisted of 6 Domains: Tool & Technology, Organization, Policy, Knowledge, Culture and Stakeholder. Further improvements for SMM, such as processes, interface, and memory analysis, were applied → November 2011.
6. Version 4.1: This version consisted of 6 domains: Tool & Technology, Organization, Policy, Knowledge, Culture, and Stakeholder. Also introduced 4 functions of SMM: firewall management, process management and denial of service detection, network detection, and port scanning for potential malware detection → December 2011.
7. Version 4.1.1 – the latest available version: This version consists of 6 domains: Tool & Technology, Organization, Policy, Knowledge, Culture, and Stakeholder. We conducted re-adjustments of parameters, re-interfacing and formulae for all ISM features. This version was installed and tested at the respondent's site for RISC investigation and SP/SQ evaluation → January 2012.

## 5.6  TESTING STAGE

We conducted the ISM deployment and testing at the respondent's site. The testing includes RISC investigation for ISO 27001 compliance and testing of the SP/SQ. RISC investigation is to find out the organization closeness over the compliance level. The SP/SQ measurement is to assess ISM performance and quality that following the eight fundamental parameters (8FPs). Moreover those measurements deal with 5-Likert scale to figure ISM performance. The highest value is "5" descript excellent performance of ISM. The lowest is "0" that disclose for not recommended performance of ISM. Those testing's deal with 10 organizations that grouped within three cluster such as ISO 27001 holder (cluster I), ISO 27001 ready to comply (Cluster II), and ISO 27001 consultant (cluster III) (Table 5.18). The testing took approximately 12 months. The details of testing stages, respondents' profile, 8FPs, and 5-Likert scale of measurement were discussed in Chapter 3 – *Methodology*.

**TABLE 5.18**  Software Testing Stage

| Respondent | Person in charge | Status |
|---|---|---|
| Banking Regulator | • Information security officer | ISO 27001 Holder |
| | • ISO 27001 officer | |
| Telecommunications | • Business portfolio and synergy data support officer | ISO 27001 Holder |
| | • ISO 27001 Implementation kick off and live test evaluation team | |
| Automotive & Manufacturing | • IT manager | ISO 27001 Ready |
| | • IT officer | |
| Financial Service | • Risk manager | ISO 27001 Ready |
| Airlines | • Scheduling operational and maintenance manager | ISO 27001 Ready |
| Health Centre | • Medical doctor | ISO 27001 Ready |
| Research Institute | • Programmer | ISO 27001 Ready |
| ICT Consultant | • System analysis | ISO 27001 Consultant |
| | • Network and security engineer | |

## 5.7 COMPARISON OF ISM WITH SOME EXISTING TOOLS

The 5S2IS, SIEM, and ICM are some available tools which assist organizations in checking their readiness for ISO 27001. The 5S2IS (5 stages to information security) consists of five major steps in assessing the readiness of an organization for compliance. Those steps are: draw up a plan, define protocols, organization measurement, monitoring, and improvement. 5S2IS is an abstraction methodology without any implementing and practical tool in the form of software to assist organizations in performing self-assessment.

SIEM (security information and event management) is a tool to conduct an analysis and pilot implementation of ISO 27001. SIEM works by analyzing the computing system, either through log data or through real time. SIEM consists of security information management (SIM) and security event management (SEM).

ICM (the implementation checklist method) is the method to pursue with the parameters of the standard. The main purposes of ICM are: (1) to gauge the level of compliance to ISO 27001 requirements by holding, department, and division. (2) Facilitate the provision of information for ISO 27001 implementation. (3) Serve as training material to understand the ISO 27001 requirements.

All those features respectively with the methods and tools are shown in Table 5.19. For instance, 5S2IS is the method that focuses on concepts and abstraction to handle the standard, but this method has not been implemented in a tool or software. On the other hand, SIEM can be used to capture network parameters. Note that the stability of the network and security is a key element in obtaining ISO 27001 certification. The limitation of SIEM is that non-technical parameters, such as strategic and managerial parameters are not covered. Whereas ICM describes 133 controls and users are given a checklist as a medium to control which parts have been implemented and which parts are not implemented. ICM is available in electronic format (an excel file). The user checks manually one by one, the controls chosen then explains how and why it is appropriate, which are applicable and which are not, and the reasons for such a decision.

**TABLE 5.19**  Existing Assessment Tools

| | Features and Function | Measurement tool | Monitoring tool | Controls | Note |
|---|---|---|---|---|---|
| 5S2IS Gillies, 2011 | • Maps the plan-do-check-act cycle onto a five-stage development process of adoption. | • 5S2IS is a methodology proposed to help an organization implement ISO 27001<br>• It is also supported by CMM (capability maturity model). No (computer) tool is provided with this method to measure an organization's compliance level. | | Covers 11 controls | Lack of practical implementation (tool or software) |
| SIEM Montesino, 2012 | • Analyze security event data in real time;<br>• Collect, store, analyze and report on log data for regulatory compliance and forensics.<br>• Provide a method to comply with the parameters of the standard | • Security information management (SIM).<br>• Provides the collection, reporting and analysis of log data.<br>• Support regulatory compliance reporting, internal threat management and resource access monitoring. | • Security event management (SEM).<br>• Real-time monitoring and incident management for security-related events.<br>• Processes log and event data from security devices, network devices, systems and applications in real time<br>• Provide security monitoring, event correlation and incident responses. | Covers 10 controls | Lack of RISC investigation that covered essential controls of ISO 27001 |

## KEYWORDS

- e-assessment
- e-monitoring
- security assessment management
- software performance
- software quality

# CHAPTER 6

# TESTING THE SOFTWARE: RISC INVESTIGATION AND SP/SQ MEASUREMENT

## CONTENTS

### 6.1   INTRODUCTION

Information security contributes to the success of organizations, as it gives a solid foundation to increase both efficiency and productivity. Many business organizations realize that compliance with information security standards will affect their business prospects. It is important to understand that protecting confidential information is not only a business requirement, but also an ethical and legal requirement (Susanto et al., 2012a, 2012b).

ISM provides the ability to enhance organizations beyond usual practices and offer a suitable approach to accelerate the compliance[1] process

[1] The process taken to achieve certification of the ISO 27001 standard. The process starts when the organization makes the decision to embark upon the exercise. The next stage is defining which part(s) of the organization will be covered by the standard. A part of this process will be selection of appropriate controls with respect to those outlined in the standard, with the justification for each decision recorded in a Statement of Applicability (SoA).

for information systems security. The new approach in ISM helps organizations improve their compliance processes by reducing time, enabling RISC self-assessment, SoA preparation, monitoring the network, and its ability to detect suspicious activities.

ISM allows an organization to perform an information security standardization project through a self-assessment process. ISM offers a new outlook in three distinct components: organization, standard and advancement tool, to assess and minimize security breaches.

This chapter discusses the role and performance of ISM in building information security literacy through a self-assessment process. For this purpose we conducted comprehensive testing for RISC investigation and SP/SQ measurement for ISM itself. The goal of this chapter (testing the software model) is to produce an analytical result from the case study to find out an organization's strategies, awareness, improvement and perspective associated with information security. In order to achieve this goal we emphasized features offered by ISM in providing self-assessment. Evaluation elements such as questionnaires, parameters and test bed variables were developed based on the literature studies (Eloff & Solms, 2000; Furnell & Karweni, 1999; Posthumus & Von Solms, 2004) as discussed in Chapter 3.

## 6.2   RESEARCH BACKGROUND

This section discusses the field research setting, information security awareness, and participants. The study setting was built on recent reviews of information security aspects and awareness in some organizations to provide a broader perspective and guidelines for the proposed framework, particularly to comply with the ISO 27001 standard.

### 6.2.1   STUDY SETTING

PWC suggested the business clusters with certain criteria considering attention towards information security and social engineering attacks. For this research we selected 10 organizations as respondents which were classified into 6 clusters as suggested by PWC. In this study, we have chosen key personnel in the organization as competent representatives in providing input, feedback, and advice associated with the research. Those personnel are in charge of information security related tasks in their respective organizations.

### 6.2.2 SECURITY AWARENESS

The ever changing business environment exposes new vulnerabilities in information assets. These vulnerabilities will increase the possibility of security breaches or attacks by hackers (Anderson & Moore, 2006; Dhillon, 2007). Cyber criminals keep adapting their techniques to exploit vulnerabilities, and it is also becoming more common (Aceituno, 2005; Easttom, 2012). Consequently the number and cost of security breaches appears to be rising fast (Potter & Beard, 2010).

Ungoed-Thomas (2003) estimated the total cost for loss of business and related security costs caused by the work of computer hackers and viruses was about £1 trillion. Leaks of confidential information because of them not being properly protected can result in detrimental publicity for an organization. As a consequence, business may be lost and customers will shift their money elsewhere since the organization is too risky (Anderson, 2001).

Potter and Beard (2010; 2012) indicated that 70% of IT budgets were spent after experiencing a security breach. Organizations are seeking new ways to prevent information security breaches and threats through the concept of *five principles of security*. The five components are planning, proactive, protection, prevention and pitfalls. *Five principles of security* identifies signs and flags of intruders in a network, establishes guidelines for safeguarding user names and passwords, and matches protection against physical access to information technology facilities with the security systems which are the most effective to deal with the threats.

Potter in ISBS 2012, said "The UK is under relentless cyber-attack and hacking is a rising risk to businesses. The number of security breaches we are experiencing has rocketed and as a result, the cost of security breaches is running into billions every year. Since most businesses now share data with their business partners across the supply chain, these numbers are startling and make uncomfortable reading for business leaders. Large organizations are more visible to attackers, which increases the likelihood of an attack on their IT systems."

Looking at this issue on the country scale, information security awareness and its breaches makes interesting discussion. Recently, Edward Snowden[2] in 2013 revealed that the NSA[3] performed espionage against EU[4] countries. The latest issue was when Snowden's leaks revealed that Australia had wiretapped high officials in Indonesia, including the President himself (Kompas.com, 2013).

"Since news broke reports of US and Australia tapping on many countries, including Indonesia, we have expressed our strong protest. Foreign Minister and government officials have taken effective diplomatic measures, while demanding clarification from the US and Australia. Today (18 November 2013) I instructed Minister Marty Natalegawa to recall Indonesia's ambassador to Australia. This is a firm diplomatic response. Indonesia also demands Australia for an official response, one that can be understood by the public, on the tapping on Indonesia. We will also review a number of bilateral cooperation agendas as a consequence of this hurtful action by Australia. These US and Australian actions have certainly damaged the strategic partnerships with Indonesia, as fellow democracies. I also regret the statement of Australian Prime Minister that belittled this tapping matter on Indonesia, without any remorse (Susilo Bambang Yudhoyono – Indonesian's President) (Detik.com, 2013)."

In Indonesia's case, ICT investment is increasing exponentially. The ICT market was indicated to be USD 15 billion in 2013 (Tribunnews.com, 2013). Consumer spending on ICT products was approximately USD 6.68 billion in September 2013 and around USD 6.81 billion in June 2013 (FinanceToday, 2013). In the field of Internet social media, Indonesia has the fourth largest number of Twitter users worldwide, with more than 4 million Twitter accounts. There are 44 million Facebook users, also ranking fourth place worldwide. The Facebook penetration in Indonesia is 18.21% of the country's population (Bakers, 2011; 2013).

[2] He is an American computer specialist, a former Central Intelligence Agency (CIA) employee, former National Security Agency (NSA) contractor. He came to international attention when he disclosed a large number of classified NSA documents to several media outlets. The leaked documents revealed operational details of a global surveillance apparatus run by the NSA and other members of the Five Eyes alliance, along with numerous commercial and international partners. ("Interview with Whistleblower Edward Snowden on Global Spying". Der Spiegel. July 8, 2013).

[3] The National Security Agency (NSA) is the main producer and manager of signals intelligence for the United States. NSA leads in cryptology that encompasses both Signals Intelligence (SIGINT) and Information Assurance (IA) products and services, and enables Computer Network Operations (CNO) in order to gain a decisive advantage for the US and its allies.

[4] European Union.

Looking at the figures above, Indonesia is a very large adopter of ICT technology. It indicates that Indonesia has a very lucrative ICT market for investors. It brings a promise for further innovation and adoption of information security technology and compliance. This compliance aims to protect organizations' assets, both tangible and intangible. Those assets, such as customer data, transactions and accounts, must be ensured that their information security is at a high level, meeting the market information security requirements and having zero tolerance for any information security breaches by following the standard, ISO 27001.

### 6.2.3 DESCRIPTION OF RESPONDENTS

In this study, systematic and selected sampling was the chosen sampling method instead of random sampling. This method is also called the *N-th name selection technique* (Patton, 2005). Respondents for this research were selectively chosen based on specific considerations and purpose.

Respondents for this study were selected with several criteria to undertake assessment. Respondents had to have an interest in compliance of ISO 27001. They had to be willing to actively participate in the whole process of the testing. The main reasons that led us to the above sampling method are due to the following criteria: *first*, the organizations have paid serious attention to information security. Either the information related to their business processes, customers or their intellectual property rights with respect to products or service they have provided.

Second, the organizations must be willing to share information associated with their information security strategies, management, ISO 27001 compliance stage, and procedural mistakes that could lead to information security breaches.

Third, the organizations must not have any objection to install and run the system (ISM) in their networks. This is an important issue as the ISM installation might affect the performance of their computer networks. ISM testing consumes a lot of time since this testing includes RISC investigation and SP/SQ measurement.

Key personnel(s) are representatives that are competent in providing input, feedback, and advice on behalf of the organizations. Those respondents are employees whose main jobs are directly related to information

**TABLE 6.1** Person In-Charge At the Organization Respondents

| Respondent's Business Sector | Person in charge | Level |
|---|---|---|
| Automotive & Manufacturing | IT manager | Middle technical level |
| | IT officer | Middle management level |
| Banking Regulator | Information security officer | Senior technical level |
| | ISO 27001 officer | |
| Financial Service | Risk manager | Middle management level |
| Telecommunications | Business portfolio and syenergy data support officer | Middle management level |
| | ISO 27001 implementation kick off and live test evaluation team | Middle technical level |
| Airlines | Scheduling, operational and maintenance manager | Senior management level |
| ICT Consultant | System analysis | Senior management level |
| | Network and Security engineer | Middle technical level |
| Health Centre | Medical doctor | Middle management level |
| | | Middle technical level |
| Research Institute | Programmer/Developer | Senior management |

security tasks and they have strong access to organizational information resources and huge responsibility in protecting information against unauthorized users (hackers). The job titles of respective respondents covered technical levels to managerial levels (Figure 6.1; Table 6.1).

## 6.3 STUDY IN BRIEF

This section briefly describes and explains the components and scenarios of the system development. The research roadmap of the whole study combines focus group discussion (FGC), software development methodologies (SDM), RISC (readiness and information security capabilities) and software performance and software quality (SP/SQ) as quantitative measurement of information security conditions and software performance.

This roadmap describes the size of respondents, procedures for data collection, techniques to analyze and derive the data for development of the framework and software, and also testing of the software in respondents' sites. The study went through the following stages: literature analysis, proposed framework based on literature study, developing a full version software, followed by testing of the software.

*Literature survey* is an analysis of recent reviews of frameworks and tools that have been developed for information security. Relevant journal searches were conducted. We conducted a thematic analysis of articles and papers associated with information security.

*Proposed framework* was the development of the new framework based on an observation of the 9STAF and respective points of concern, advantages, and disadvantages. The 9STAF were studied carefully and a comparison among them was carried out to reveal their salient features and respective positions. Once the proposed framework was developed, user requirements were determined to develop software as a tool that can help an organization to better position itself for ISO 27001 compliance.

*Software development* was the translation and interpretation of ISF that had been incorporated with ISO 27001 components into a software form, ISM, to help users review their information security status and circumstances.

*Software deployment and testing* consisted of the RISC investigation for compliance and obtaining the SP/SQ measurements. The respondents were divided into three clusters: ISO 27001 holders (cluster-I), ISO 27001 ready (cluster-II), and ISO 27001 consultants (cluster-III).

## 6.4 RESULTS

The results of the study revealed that the majority of organizations recognize the importance of information security governance as an integral factor for the success of ICT and corporate governance. Most respondents had clear information security strategies or written information security policy statements. The ISO 27001 holders had disaster recovery plans to deal with information security incidents and emergencies. They clearly defined and communicated information security roles and responsibilities. The results show that alignment between information security governance and

**FIGURE 6.1**    Field study for RISC investigation and SP/SQ measurement.

the organization's overall business strategy is adequately implemented. The results also showed that risk assessment procedures were adequately and effectively implemented. For software performance measurement the ISM scored 2.7 out of 4, meaning highly recommended to use it for self-assessment and further SoA preparation.

## 6.4.1  PART-I: INFORMATION SECURITY STRATEGIES

The analysis below emphasizes on the lessons learnt from testing the system. This section attempts to discuss the learning curve, challenges, barriers and experiences while brainstorming and testing ISM from the perspective of the respondents who were ISO 27001 holders.

Both organizations have vast experience in compliance processes. Based on their experiences, BI and Telkom are eligible to provide objective opinions dealing with performance and function of ISM. The findings are important to trigger further research in a broader context of security compliances in the organizations. These opinions and feedbacks will be discussed in Subsection 6.4.4.4 – *Enhance Security Literacy.*

### 6.4.1.1  Bank of Indonesia: A Success Story

The ISO 27001 certification process in the Bank of Indonesia (BI) monetary authority started from the division of information and communications technology (ICT) in 2007. The aim of compliance was to standardize and control the workflow and processes of banking financial transactions. The employees involved in this process were called security officer teams. Twenty employees were assigned to support the implementation of the information security tripartite: confidentiality[5], integrity[6], and authority[7].

The main responsibilities of the security officer teams are detailed further in the following points:

1.  Register all hardware and software connected to the organization network to ensure that all components involved in the network are well-monitored, particularly to prevent information security breaches.
2.  Regularly update usernames and passwords every maximum period of three months.

---

[5] Confidentiality refers to preventing the disclosure of information to unauthorized individuals or systems. For example, a credit card transaction on the Internet requires the credit card number to be transmitted from the buyer to the merchant and from the merchant to a transaction processing network.
[6] Integrity means maintaining and assuring the accuracy and consistency of data over its entire life-cycle. This means that data cannot be modified in an unauthorized or undetected manner.
[7] Availability means that the computing systems used to store and process the information, the security controls used to protect, and the communication channels used to access must be functioning correctly.

3. Strengthen the network with firewalls, traffic monitoring, process management and port management, to ensure that the network is running properly and provides early prevention against intruders, hackers, and unauthorized users.

4. Create an imaginary line (blue line) that encompasses the ISO 27001 zone. No-one can pass inside the line without permission from the security officer. Any unauthorized personnel who enters the imaginary line area is identified as a security intruder, and the event is classified as a security incident.

5. Provide the log activity documentation to capture employee and guest activity within the ISO 27001-zone. This document includes names, origins and destinations in the zone. Log times of entry and exit of the zone should also be well-documented to monitor their activities, in case any security breaches occur while inside the zone.

6. Preparing data and information backup related to the business processes. This backup is recorded in a storage medium such as a compact disc, external hard disk, USB drive, or filing cabinet folders. The original data and its backup should not be stored in the same room area, following disaster recovery system scenarios.

In BI's case, the adoption process for certification took approximately 3 years. It was assisted by consultants with a considerable cost. In this stage, the security officer team constructs an entire listing of business processes. This list described all stages of business processes together with potential security breaches. Once potential security breaches defined, worst-case and best-case scenarios were constructed to overcome those security breaches. These scenario documents were used as reference to propose the SoA document.

Assisted by LEMTIUI[8] as the information security consultant, BI has successfully obtained certification of ISO 27001 since 2008 from Bureau Veritas Quality International (BVQI).

*"LEMTIUI has successfully supported Bank of Indonesia in implementing ISO 27001. The strong commitment and effort by LEMTIUI and BI have resulted in an ISO 27001 certificate for BI in 2008. The compliance was successfully awarded by the Bureau Veritas certification body.*

---

[8] ICT information security consultant subsidiary of the Department of Industrial Engineering, University of Indonesia.

*BI is the second central bank in the world that has been awarded with ISO 27001 certification after New York Federal Reserve Bank. The ISO 27001 auditors, Ganesh from India (audited during 2006–2007) and Joe from Taiwan (audited during 2008) were impressed by the commitment and implementation process by BI staff and management"* (lemtiui.com).

Once ISO 27001 was certified, the next task is to maintain their security performance. In the first six months after certification, many data and documents had to be securely protected. There were approximately 100 bundles of documents daily, each bundle consisting of more than 100 titles. In this early period, there were many information security breaches conducted by the staff themselves, especially those who do not really understand the concept of organization information security. With passion and continuous socialization and education of employees, information security violations began to decrease significantly as seen in Figure 6.2.

When the interview took place in 2012, the ISO 27001 maintenance team involved approximately 200-300 employees from various levels of organization. This team's objective was to keep the spirit of information security, since the certification is only valid for a period of 3-5 years. The standards body will perform re-inspection in the organization to make sure that the security controls still comply with the ISO 27001. Here, ISM could grab their necessity for a tool to review and accelerate the security-update and SoA construction.

## 6.4.1.2  Telkom Indonesia Corp: Lessons Learned

Telkom Indonesia is a company listed in the stock exchange in several countries. Telkom is identified in Indonesia Stock Exchange (code: TLKM), London Stock Exchange (code: TKID), and New York Stock Exchange (code: TLK). It is a provider of information and communication as well as telecommunication services. Telkom as the largest telecommunications company in Indonesia has a huge number of customers. It has fixed-line subscribers numbering around 30 million and more than 150 million subscribers for mobile phone services. Nowadays, the corporation covers a number of subsidiary companies in the area of information, telecommunication, and satellites. Those subsidiary companies include Telkomsel, Multimedia Nusantara, Infomedia, Telkom Vision, TelkomFlexi,

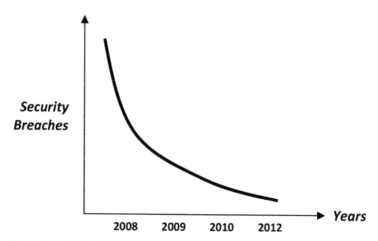

**FIGURE 6.2** Information security violation by internal staff.

TelkomSpeedy, TelkomAkses, Telkomsel Flash, simPATI, TelkomNet Instant, and Pacific Satelit Nusantara (Telkom Satellite-1, Telkom Satellite-2, and Telkom Satellite-3)[9].

The organization's aims for ISO 27001 compliance was to provide products of IT services, Internet Protocol connectivity and Virtual Personal Networks that satisfy the security standard. TUV Rheinland[10] is the standards body which awarded the certificate to Telkom Indonesia Corp (Figures 6.3 and 6.4).

The organization's compliance processes, called *implementation commencement* consisted of internal quality audit (IQA) and external quality audit (EQA) stages. To meet with the ISO 27001 requirements, Telkom highlighted three information scenarios as implementations of security controls. Those scenarios are: infrastructure management program, disaster recovery system, operation and maintenance. The scenarios will be discussed in the following subsection.

---

[9] Telkom Annual Report and obtained from www.telkom.co.id
[10] TÜV Rheinland is a global provider of technical, safety, and certification services. Originally called the Dampfkessel-Überwachungs-Verein (Steam Boiler Inspectorate), TÜV Rheinland was founded in 1872 and has its headquarters in Cologne, Germany.

**FIGURE 6.3**  Field study for RISC investigation and SP/SQ measurement.

**FIGURE 6.4**  ISO 27001 certificate awarded to Telkom Indonesia Corp.

### 6.4.1.2.1  Infrastructure Management Program

An infrastructure management program (IMP) aims to set up and construct the organization's infrastructure to follow the ISO 27001 infrastructure safety scenario. The IMP tasks are described as follows (Table 6.2):

The main component of IMP is standardization of items. The purpose of this is to ensure that every infrastructure component follows the rules of the ISO 27001 standard. The details of standardization of items are explained as follows:

1.  The active device(s) is labeled to indicate its function and configuration within the corporate network (backbone) (Figure 6.5).
2.  Employees who enter an active device area have to be recorded in its activity log book.

**TABLE 6.2**   ISO 27001 Tasks and Corporate Actions

| No | ISO 27001 Task | Corporate Actions |
|---|---|---|
| 1 | Construct the scenarios and functions related to information security controls. | 1. Create a function that carries out activities related to information security.<br><br>2. Construct job descriptions for these functions. |
| 2. | Inventory of assets. | 1. Collecting data on the assets associated with the targeted scope of certification.<br><br>2. Perform a classification of assets. |
|  | Deliverable | The list of assets and classification document. |
| 3 | Perform risk analysis (risk | 1. Identification of risks.<br><br>2. Classification of risk.<br><br>3. Determination of the strategy in the face of risk and its impact.<br><br>4. Determination of the impact to risk acceptance levels. |
|  | Deliverable | All documents associated with risk analysis, which are required by ISO 27001 |
| 4 | Define objectives and steps to achieve information security targets. | 1. Define goals of information security.<br><br>2. Revise and evaluate controls that required by ISO 27001. |
|  | Deliverable | 1. Objectives of information security.<br><br>2. All required documents related to the rules and controls of ISO27001. |
| 5 | Determine the measurement method for measuring effectiveness and readiness level of controls and clauses. | 1. Define and determine effectiveness level of the control's application.<br><br>2. Define and determine the method for measuring effectiveness level of the control's application.<br><br>3. Measuring and assessing effectiveness level of the control's application. |
|  | note:<br>possible to conduct by ISM for RISC investigation |  |
|  | Deliverable | Prepared records of control measurement results. |
| 6 | Define the statement of applicability (SoA). | Createa SoA document in accordance with the requirements of ISO27001. |
|  | Deliverable | SoA document. |

**TABLE 6.2** (Continued)

| No | ISO 27001 Task | Corporate Actions |
|---|---|---|
| 7 | Prepare the Business Continuity Plan (BCP) document for ISMS implementation | Develop BCP documents, assign teams, BCP scenario and testing all those things. |
|   | Deliverable | 1. BCP document and testing scenarios. |
|   |   | 2. BCP testing result. |
| 8 | Assisting the ISMS implementation | ISMS implementation of the management process through regular review, internal audit, and management review. |
|   | Deliverable | ISMS management reports. |

3. Employees/outsourcing partners/guests are all issued identity cards.

4. The active devices rooms shall be provided and equipped with the following label (Figure 6.6).

5. Monitoring cameras should be installed for real-time monitoring in the active devices room (Figure 6.7).

6. The active device room has to be locked and the key placed securely in a locked box (Figure 6.8).

7. Each desktop or laptop has to be recorded and protected with an alphanumeric password at least 8 characters long with a maximum 5-minute password-protected screen saver setting.

8. Any outsourcing partner and vendor who works for and on behalf of the corporation shall sign a non-disclosure statement.

9. All active devices must be well recorded with scheduled maintenances. If there is any maintenance delay, the justification for such a delay must be written and recorded properly.

10. Employees who access any active device must pay attention to his/her personal access password and change their login-password authority regularly as mentioned in the SOP[11].

11. It is prohibited to display the customer identity (i.e., BNI, BRI, BCA, Mandiri, Permata) in active devices and network ports to avoid hacking, phishing and infrastructure sabotage by internal staff, vendors, outsourcers and other parties (Figure 6.9).

---

[11] Standard operational procedure.

**FIGURE 6.5**    Active device labels.

**FIGURE 6.6**    Safety label for active devices room.

**FIGURE 6.7**    Monitoring camera.

**FIGURE 6.8**   Key holder box.

**FIGURE 6.9**   Network port: customer identity.

12. USB flash drives are prohibited to be used within the 'safety area' of the corporate security zone (Figure 6.10).

13. The entire active devices' data and software should be backed up in a certain media (CD, hard drive, etc.) within certain periods: daily, weekly, and monthly. The backup media must be stored in a safe place (2 copies of backups each session).

### 6.4.1.2.2   Disaster Recovery System

A disaster recovery system (DRS) is the scenarios, processes, policies and procedures related to preparing for recovery or continuation of an organization's vital technology infrastructure after an incident or disaster. Disaster recovery focuses on the IT or technology systems that support business functions, as opposed to business continuity, which involves planning to keep all aspects of a business functioning in the midst of disruptive events.

DRS is a subset of a larger process known as business continuity planning and includes planning for resumption of applications, data, hardware, electronic communications (such as networking) and other ICT infrastructure. DRS is the implementation of the ISO 27001 objective (clause) to ensure the availability of business processes during any incident or disaster. That objective consists of five essential controls that specify scenarios of DRS associated with information security incident management, IT service continuity management and configuration management in handling a security incident. Those controls are: (1) information security in the business continuity process; (2) business continuity and risk assessment; (3) developing and implementing continuity plans, including information security; (4) business continuity planning framework; and (5) testing, maintaining and re-assessing the business continuity plan.

**FIGURE 6.10**   Security Zone: no USB flash drive is allowed.

To support DRS scenarios, Telkom Indonesia Corp assigned approximately 50 employees to handle the disaster related tasks which encompass planning, operation, and maintenance. The objectives of DRS planning for business processes are as follows:

1. To improve human resources competency, particularly related to the strategic roles in the operation, maintenance, control, and infrastructures of disaster planning.

2. To prepare human resources competency in the appropriate system recovery plan, minimizing disruptions from a major or minor disaster.

3. To support the Disaster Recovery Plan, which includes in it the rollback procedure from the DRS to the commercial system, to minimize revenue loss.

4. Provides a quick and appropriate scenario for the handling of major or minor disasters to prevent loss and outage. Disasters can have possible negative impact on brand image and product image if an organization does not properly handle the situation.

5. To be able to run a live test implementation, which is one of the essential requirements of ISO 27001 certification.

The DRS system is the corporate backup during incidents and disasters. On the other hand, the current running system, which is the commercial system, is the one that provides the full service and operations for the organization's business processes.

In Telkom's case, the commercial system is located in two main cities in Indonesia, namely Jakarta and Surabaya, while the DRS is located in the city of Semarang (Figure 6.11). The DRS system works by continuously synchronizing with the commercial database in real-time and peer-to-peer. In case of a commercial system crash (down) because of a disaster, the DRS system is ready to take over the commercial system by activating the commercial system replication data (data subscriber, charging and service). The scenarios of commercial system replication activation are shown in Figures 6.12–6.14. The hardware specifications of DRS can be seen in Table 6.3.

Figure 6.11 describes the DRS configuration within the corporation to handle disasters. The DRS architecture and signaling configuration diagram are shown in Figure 6.14, portraying how each DRS system synchronizes

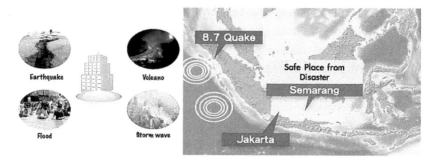

**FIGURE 6.11**   Disaster Recovery System locations.

data with the commercial system in Jakarta and Surabaya (Figure 6.16). The main reason for placing the commercial system in Jakarta and Surabaya is because the potential for disaster to occur simultaneously in both cities is very small.

*System Specification*

For services continuity, the running commercial system constantly replicates the data, such as customer data, accounts, and charging information to the DRS system, synchronizing the data during normal conditions. In

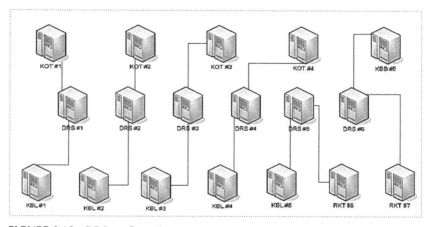

**FIGURE 6.12**   DRS configuration.

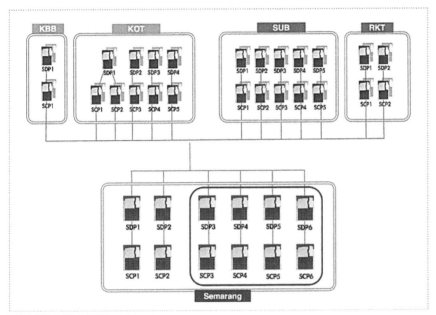

**FIGURE 6.13**    DRS architecture.

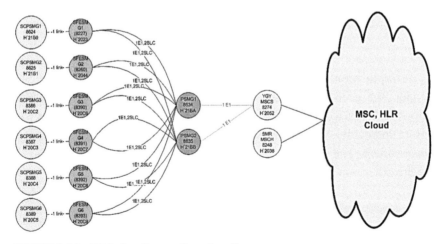

**FIGURE 6.14**    DRS signaling configuration diagram.

case a disaster happens, all data will be restored from the DRS (Figure 6.16).

**TABLE 6.3** DRS's Hardware Specification

| System | Specification |
|---|---|
| SCP | • HP server 8-way rp7440 FAST Solution |
| |    – PA-8900 1 GHz CPU 4P 8Core; 8GB Memory; 300 GB 15K HDD x 2 |
| | • HP (factory-racked) 5300 Tape Array x 1 |
| | • HP DAT 72 Array Module X 2 / DAT 72 Tape x 10 |
| | • HP Universal Rack |
| | • HP DL380 G |
| | • ADAX solution |
| |    – SS7 72Link, SIGTRAN M3UA & M2PA |
| SDP | • HP server 8-way rp7440 FAST Solution |
| |    – PA-8900 1 GHz CPU x 8; 32GB Memory; 300GB 15K HDD x 2 |
| | • HP (factory-racked) 5300 Tape Array x 1 |
| | • HP DAT 72 Array Module X 2 / DAT 72 Tape x 10 |
| |    – HP Universal Rack |
| Storage | • HP EVA4400-A System |
| |    – HP Storageworks SAN Switch x2 |
| |    – HDD: 146GB 15K RPM FC HDD x 8 |

A replication process involves two DRS components, namely the sender and receiver, through socket connection to the TCP/IP[12] backbone network. When a database server sends requests by several queries such as insert/update/delete, those queries will be automatically listed in the *log-queue query*. The sender reads the log-queue and sends them to a receiver on the other server (Figure 6.15).

### 6.4.1.2.3 Operations and Maintenance

The function of operations and maintenance (OAM) of business processes is to monitor a system according to a DRS scenario. The OAM consists of business

---

[12] The transmission Control Protocol (TCP) and the Internet Protocol (IP). TCP/IP provides end-to-end connectivity specifying how data should be formatted, addressed, transmitted, routed and received at the destination. This functionality has been organised into four abstraction layers which are used to sort all related protocols according to the scope of networking involved.

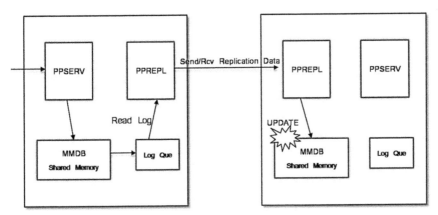

**FIGURE 6.15** DRS replication scenario.

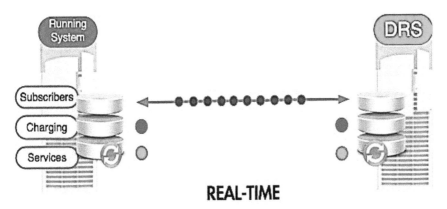

**REAL-TIME**

**FIGURE 6.16** DRS real time replication.

process components such as regular servers, DRS servers, synchronization, and database hardware.

There are two main tools to support OAM processes, advanced OAM system (OAMS) and dynamic CRM (DCRM). The minimal hardware requirements to run OAMS and DCRM are NIC-1 EA and dual monitors. The recommended operating system is Windows 7 Family (other OS's will not be supported) and Microsoft.NET Framework v.4.0.

OAMS has a platform with the user interfaces as indicated in Figures 6.17–6.19. The details of OAMS features are as follows:

1. The dashboard application can customize monitoring and layout according to user requirements.

2. The Topology maps feature provides real-time monitoring according to the user's needs, supported by mapping shapes (line, rectangle, round rectangle, ellipse), images (BMP, JPG, GIF, PNG), text and monitoring objects (server, channel, real-time graph, custom).

3. Alarm management for suspect activities.

4. Additional features for statistics analysis.

The DCRM involves three departments: marketing, sales and service (Figure 6.16). The advantages of DCRM are (1) supports different interface access such as outlook, web, and mobile. (2) DCRM could be integrated with Microsoft Office (Outlook, Excel). (3) High flexibility and low maintenance costs. (4) Supports the dot NET framework. (5) Adaptable, meaning that it can be easily and quickly adapted for future upgrade of emerging technology.

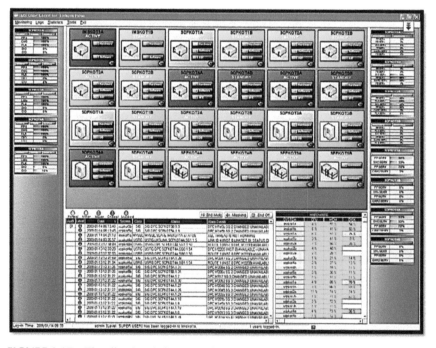

**FIGURE 6.17**   The client's existing operation and monitoring tool.

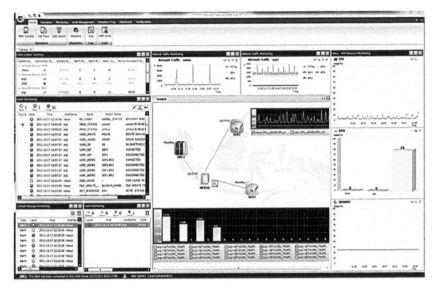

**FIGURE 6.18**   Advanced operation and monitoring tools.

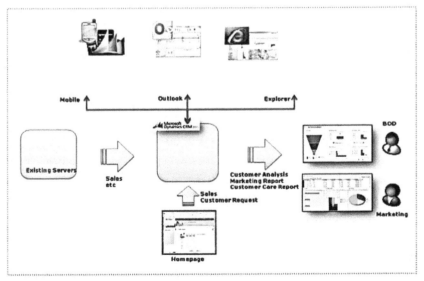

**FIGURE 6.19**   System configuration by dynamic CRM.

## 6.4.2   PART II: INFORMATION SECURITY BEHAVIOR

In the quantitative phase (Part II) of the research stage, we conducted surveys and interviews with the respondents through semi-structured questionnaires, which are used to obtain input and feedback. The questionnaire contains three topics, namely: attitudes to information security, security awareness, and security standards.

### 6.4.2.1   Attitudes to Information Security

The ever-changing business environment increases the challenge in protecting information assets. There are three strategies that need to be integrated in addressing security issues. The first is a security awareness program. The main objective of this program is to educate and train employees that information security is a serious matter. Employees must have basic knowledge in information security and know how to avoid security problems and respond appropriately if threats occur.

The second strategy is backup and recovery of any data and information as any security prevention method cannot guarantee 100% security. In case of any security incidents, the loss of data and information is minimal or insignificant as all data and information could be recovered. The third strategy is to strengthen information system security through compliance with an information security standard such as ISO 27001. This may imply changes in policy and procedures in dealing with security issues as dictated by the standard.

The third strategy normally includes the first and the second strategies. Therefore, compliance with ISO 27001 will help organizations to address their information security issues, which will lead to improve confidence in dealing with information security problems and customer trust.

Respondents claim that protecting customer information remains the most important driving force for security. Preventing downtime, outages, protecting the organization's reputation, and maintaining data integrity are important issues. Therefore, complying with the standard is the biggest driver of their organization's information security strategies.

It seems respondents were aware of the ISO 27001 information security standard. Respondents have implemented ISO 27001 since

2012 (Telkom Indonesia Corp) and 2008 (Bank of Indonesia), and the remaining are undergoing processes for compliance. However, a security policy is only useful if the staff understand and adhere to it.

Respondents state that the use of electronic activities through the Internet such as electronic commerce, electronic portfolio, electronic office and electronic meeting for supporting business processes is necessary. Several uses of the Internet to support corporate business processes in order of priority are as follows (Figure 6.20): (1) employee web access; (2) e-procurement and electronic data interchange; (3) transactional website; (4) the company website profile and general information services; (5) the use of internet e-mail to support business processes, especially for the employee communication.

### 6.4.2.2  Security Awareness

According to an information security breaches survey (ISBS, 2010), 31% of small organizations and 41% of large organizations are aware of ISO 27001 and 30% of them have fully implemented the standard. The same conclusions were derived from Ernst and Young (2010) who stated that 8% have achieved formal certification and 36% of those were using ISO 27001 as a basis for their with information security management system

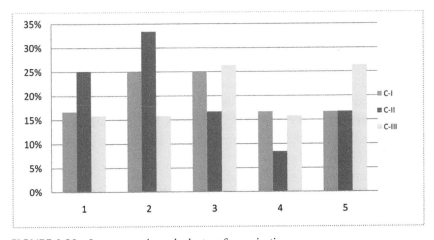

**FIGURE 6.20**   Internet use in each cluster of organizations.

(ISMS). In this study, 20% of respondents have fully implemented ISO 27001, and 70% were planning to implement ISO 27001 and preparing the required documents for compliance with ISO 27001. Some surveys such as ISBS, 2008, 2010 and 2012, also including this study, have indicated that ISO 27001 is a widely accepted information security standard and awareness rates are high.

The business environment is changing rapidly. Organizations are becoming increasingly interconnected through networks. The changing business environment will lead to increasing challenges for protecting information (Potter, 2010). Since these changes create new vulnerabilities, criminals are adapting their techniques to exploit the vulnerabilities; as a result, cybercrimes are becoming more common. These cybercrime attacks are creating a demand for information assurance and mandatory security requirements are becoming more common. More organizations need to comply security standards. It is encouraging that the number of organizations with a formal security policy, such as ISO 27001, is high.

Despite the global economic downturn during 2012–2013, the average IT budget for information security continues to grow, particularly in companies involved in this study. In our case, the organizations' objectives through allocating budget in information security were (Figure 6.18a): (1) to protect customer information; (2) to comply with laws set by the regulator; (3) maintaining the reputation of the organization; (4) maintain data integrity and ensure business continuity in unexpected situations (disasters); and (5) improve efficiency and protect other assets.

Respondents also seem to spend more budget in response to serious security incidents. Around 40% of respondents had suffered a serious incident such as breach of data protection laws or regulations, misuse of confidential information, loss or leakage of confidential information and unauthorized access to systems or data. The seriousness of incidents experienced appears to have impact on decisions taken to increase security expenditure. The respondents' information security budget distribution are as follows (Figure 6.21a):
1.  No information security budget (0%).
2.  Less than 1% of IT budget (12.5%).
3.  Between 2% – 10% of IT budget (25%).

4. Between 11% – 25% of IT budget (25%).
5. More than 25% of IT budget (37.5%).

Those figures show that the average expenditure for information security is now nearly 25% of IT budget. It is the highest level of information security expenditure compared to the ISBS survey of 2010, 2012, that
showed that it was around 10% of IT budgets.

However, respondents remain, pessimistic about what the future holds.
Respondents expect to have more security breaches in the future and it
will be more difficult to identify the type of security attacks. There are
alternative solutions to resolve those security breaches (as shown in Figure
6.21*b*); *first,* have better training and awareness programs for employees
and stakeholders to remind and make them aware that information security
is important. *Second,* make amendments to contingency plans and provide
data backups to support business processes during incident and disasters.
*Third,* improve the security policy particularly associated with the information security standard, ISO 27001.

As consequence, respondents state that information security is a high priority issue to base the board of management's decision making. The security
improvements and scenarios have been prepared to prevent similar future incidents. Financial services, bank regulator, healthcare, airlines and technology

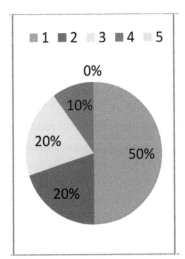
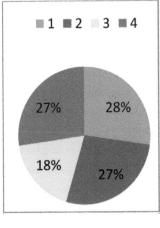

**FIGURE 6.21**   (a) Information security budget. (b) Alternative solution for future
incident.

organizations assign the highest priority to security; in this study no sectors indicated that their information security aspect is a low priority.
Security Strategy and Controls

### 6.4.2.3  Security Standard

Saint-Germain (2005) argued that an important factor for ISO 27001 certification is to demonstrate to partners that the organization has identified, measured and mitigated their security risks. Currently several standardization organizations exist at various levels: international, regional, and national level. Organizations which have published information security standards that have gained wide acceptance are ISO, Information Systems Audit and Control Association (ISACA), Information Systems Security Association (ISSA), National Institute of Standards and Technology (NIST), British Standards Institution (BSI), Information Security Forum (ISF), and Payment Card Industry Security Standards Council.

Several security standards are continuously published and gaining acceptance. Some of these standards provide guidelines, others promote best practices. However, only a few of them provide a basis for certification. There are five well-recognized information security standards in the world (Susanto et al., 2011a, 2011b), namely ISO 27001 (International Standard Organization), COBIT (Control Objectives for Information and Related Technology), ITIL (Information technology infrastructure library), PCIDSS (payment card industry data security for standard), and BS7799 (the British Standard). In this study, respondents mentioned three information security standards that have been adopted as their security policy guideline. Those standards are ISO 27001 (70%), COBIT (15%), and ITIL (15%). ISO 27001 is clearly the most widely accepted and recognized as a good standard by organizations.

The main challenges and barriers faced by respondents in the process of adoption of a standard is to understand the terms and concept (40%), preparing documents (40%), and the allocation of a huge cost (20%). These difficulties prolong the certification process. For example, only 25% of the companies surveyed can comply with ISO 27001 within a period of 6 to 12 months. The rest needed more time: 1-2 years (60%), or more than 2 years (15%).

The respondent organizations normally employ a consultant to help them comply with ISO 27001. The cost for the ISO 27001 compliance process varied between IDR 300 million to IDR 2.5 billion ($33,000 to $250,000) depending on the scope and number of branches included. Whilst this figure has limited relevance due to respondent size, it does show that cost can be a barrier for adoption. Fomin et al. (2008) used the longer experience of ISO 9001 and ISO 14001 as references to consider the barriers to adoption of ISO 27001. They argued that the benefits of ISO 9001 certification have gradually shifted from earlier times when its certification was used as a positive signal and selling point to markets and customers where firms can actually gain direct benefits from the effective use of the quality management system.

Cherdantseva & Hilton (2013) stated that ISO 27001 is becoming the lingua franca for information security. Potter and Beard (2012) revealed that ISO 27001 is enjoying greater acceptance and its adoption grows 15% annually. The increasing adoption of ISO 27001 is also evident from the growing number of certifications world-wide. The ISO surveys of 2008 reported that ISO 27001 certifications are increasing steadily; certifications have increased by approximately 20% per annum.

### 6.4.3 PART III: RISC INVESTIGATION

ISM deployment and testing were conducted in the respondents' site. The testing includes RISC investigation for ISO 27001 compliance. There are three clusters for the testing: cluster-I are ISO 27001 holders, cluster-II consists of organizations pursuing ISO 27001 certification, and cluster-III are the ICT information security consultants. RISC investigation involved 6 domains, 7 clauses, 21 controls, 144 assessment issues and 256 refined questions.

The 5-point Likert scale was used to conduct the RISC investigation. The scale describes the implementation level of controls that comply with ISO 27001, which are described as follows; "0" = not implemented, "1" = below average, "2" = average, "3" = above average, "4" = excellent.

## 6.4.3.1   Respondents' Self-Assessment

In this section we provide an example of RISC assessment for compliance conducted by respondents to reveal their organization's information security circumstances compared to ISO 27001 requirements.

The results given are illustrated in the following figures' description. Figure 6.22 summarizes the results of all domains together with their associated controls described in ISF. Figure 6.24 is given to illustrate the strengths and weaknesses in the performances of each control.

Figure 6.23 depicts the achievement conditions of the six domains of ISF together using histograms. Figure 6.24 shows a condition for the domain "Tool & Technology". Figure 6.25 indicates the 21 essential controls by histograms. The overall score of all domains is "2.66" (see Table 6.4). The score for the domain "policy" is 4, the highest, which means that an organization has excellently satisfied the related control, namely "information security policy document".

The domain "knowledge" has three controls, namely; (1) intellectual property rights, (2) protection of organizational records, (3) data protection and privacy of personal information. The "knowledge" domain achieved a grade of

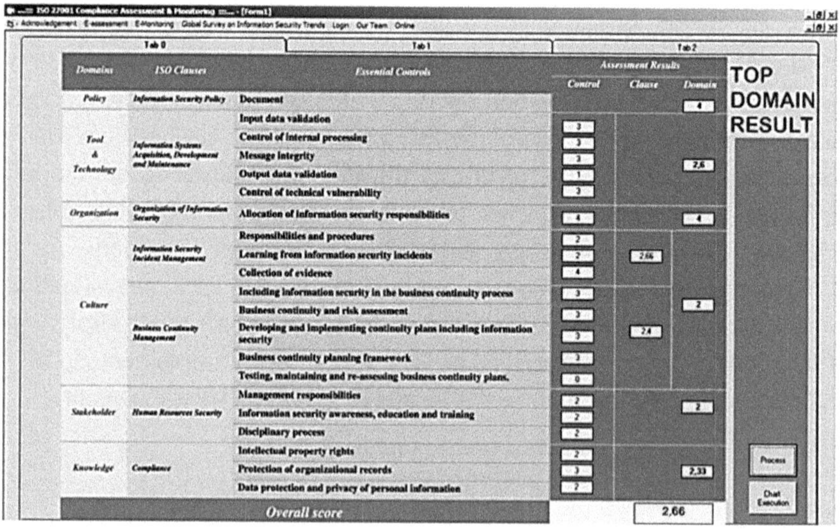

FIGURE 6.22   Final result view on summarize style.

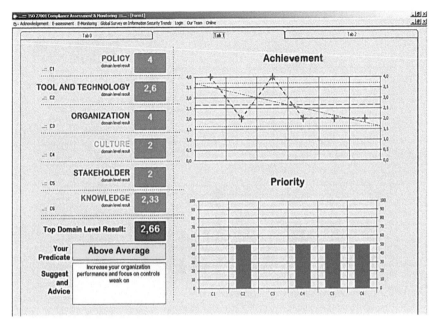

**FIGURE 6.23** Achievement and priority result at the "Tools and Technology" domain level.

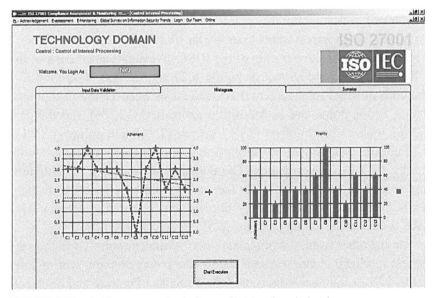

**FIGURE 6.24** Achievement and priority result at top domain level.

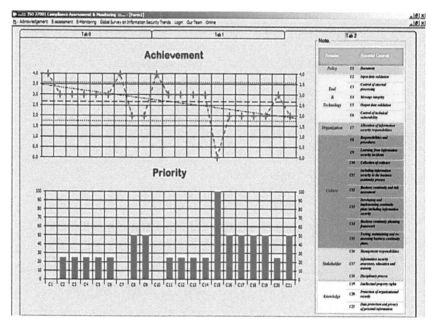

**FIGURE 6.25**    21-essential controls final result at top domain level

2 (average), indicating that the organization achieved "average performance" and needs further improvement to satisfy the ISO 27001 requirements.

The comprehensive results of the RISC investigation for each of the 10 organizations are shown in Figure 6.26 and 6.27(a). There are five domains of RISC achievement that are the respondents' strongest points. Those strong points are as follows; organization (12.5%), stakeholders (25%), tools and technology (25%), culture (25%), and policy (12.5%).

Figure 6.27(*b*) shows the "strong point" domains with the highest score during the RISC investigation. The domains organization, stakeholders, and culture, were each scored as the strong point by 25% of the respondents, while the tool-technology domain and knowledge domain were each scored as the strong point by 12.5% of respondents.

On the other hand, the organization's "weakness point", which is the lowest scoring domain, was indicated as the policy domain, scoring lowest in 50% of the respondents, and the domains of culture and organization, each scoring lowest in 25% of the respondents (Figure 6.27 (*c*)).

**TABLE 6.4** An illustration of RISC Investigation

| Domain | Clauses | Controls | Organization's Assessment Result | | |
|---|---|---|---|---|---|
| | | | Controls | Clauses | Domain |
| Policy | Information Security Policy | Document | 4 | 4 | 4 |
| Tool and Technology | Information Systems Acquisition, Development and Maintenance | Input data validation | 3 | 2.6 | 2.6 |
| | | Control of internal processing | 3 | | |
| | | Message integrity | 3 | | |
| | | Output data validation | 1 | | |
| | | Control of technical vulnerability | 3 | | |
| Organization | Organization of Information Security | Allocation of information security responsibilities | 4 | 4 | 4 |
| Culture | Information Security Incident Management | Responsibilities and procedures | 2 | 2.67 | 2.84 |
| | | Learning from information security incidents | 2 | | |
| | | Collection of evidence | 4 | | |
| | Business Continuity Management | Including information security in the business continuity process | 3 | 3 | |
| | | Business continuity and risk assessment | 3 | | |
| | | Developing and implementing continuity plans including information security | 3 | | |
| | | Business continuity planning framework | 3 | | |
| | | Testing, maintaining and re-assessing business continuity plans | 3 | | |

**TABLE 6.4**  (Continued)

| Domain | Clauses | Controls | Organization's Assessment Result | | |
|---|---|---|---|---|---|
| | | | Controls | Clauses | Domain |
| Stakeholder | Human Resources Security | Management responsibilities | 2 | 2 | 2 |
| | | Information security awareness, education and training | 2 | | |
| | | Disciplinary process | 2 | | |
| Knowledge | Compliance | Intellectual property right | 2 | 2.33 | 2.33 |
| | | Protection of organization records | 3 | | |
| | | Data protection and privacy of personal information | 2 | | |
| | | Overall score | | | **2.66** |

0 = not implementing; 1 = below average; 2 = average; 3 = above average; 4 = excellent.

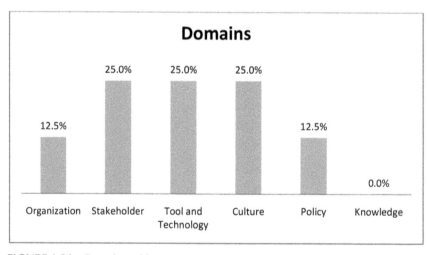

**FIGURE 6.26**  Domains achievement.

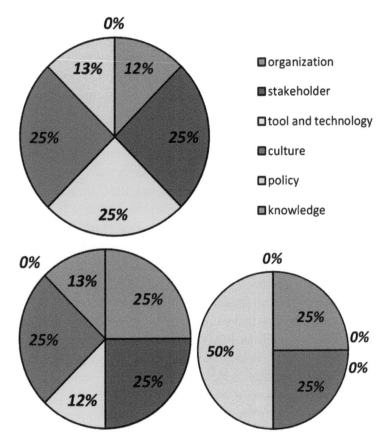

**FIGURE 6.27**   An ISMS illustrative measurement.

## 6.4.3.2   Information Security Self-Awareness

The most important function of ISM is that it provides the ability for organizations to participate, forecast, and actively assess their own information security circumstances. Self-awareness is the one of key indicators for improving readiness and capabilities of information security aspects. The respondents appreciate that this aspect provided by ISM will improve their understanding and knowledge of security and investigation stages. From the interview session, we found that ISM truly provided users with the abilities of self-investigation and real-time monitoring of network activities.

The users from cluster-I, cluster-II, and cluster-III explained how they gained benefit from ISM in terms of security knowledge. ISM helps them identify their weak controls quickly based on the information provided by the achievement-priority feature. Users also said that ISM can significantly reduce the time to produce the SoA document during the security compliance process.

The level of a user's literacy on information security is indicated by the time he/she spent on conducting RISC measurement using ISM. The results of the information security literacy survey are very interesting. The average time spent by users for conducting RISC investigation using ISM is around 720 min or 12 h (Figure 6.28). This result reflects that ISM highly assists the users by decreasing the duration of RISC investigation. It revealed that the interactions with ISM have improved the users' information security self-awareness and literacy. This survey also revealed that ISM can provide information in a clean interface to increase their security knowledge and abilities to support their RISC investigation.

The above facts show that there is a vast difference between ISM and current approaches such as ICM[13]. On average, the respondents have improved their security literacy and successfully improved the organization's information security self-awareness after interactions with ISM. Compare ISM's results to ICM, using which Bank of Indonesia took around 12 months, and Telkom Indonesia approximately 6 months. This is in line with Kosutic (2010, 2013) who stated that organizations normally take 3-36 months to complete the RISC investigation and compliance processes (see Table 6.5).

Users from Cluster-I said that ISM can possibly be used to monitor current circumstances associated with information security issues, including the infrastructure management, recovery system, worst case scenarios, and risk management. Users from Cluster-II stated that ISM can provide a semi-automated assessment for compliance level, before inviting assessors for ISO certification. Users from Cluster-III said that ISM will help consultants to provide systematic advice to companies which are interested to pursue ISO 27001 certification processes, rather than to use the current checklist form (either hard-copy or soft-copy format).

---

[13] Implementation Checklist Method.

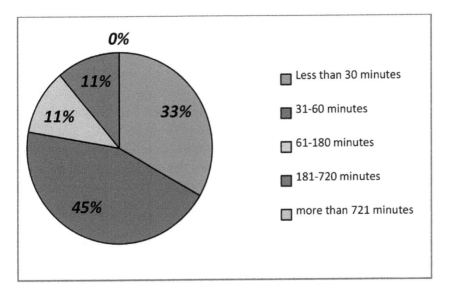

**FIGURE 6.28**   An ISMS illustrative measurement

**TABLE 6.5**   A Comparative for RISC Investigation Duration: ICM and ISM

| Organizations | ICM | ISM |
|---|---|---|
| Bank of Indonesia | • 36 months; starting from SoA preparation until certification.<br>• 12 months for RISC investigation | • Approximately 12 h for RISC investigation |
| Telkom Indonesia | • 12 months; starting from SoA preparation until certification.<br>• 6 months for RISC investigation | • Approximately 12 h for RISC investigation |
| Cluster-II | • 24-36 months*; starting from document preparation until certification. | • Approximately 5–12 h for RISC investigation |

*average time and respondents expectation based on current compliance standards; COBIT, ITIL, Microsoft, CISCO, National Standard.

## 6.4.3.3   Management Support

Management support is the most important factor for the success of compliance processes. In this study, the board of management of Telkom Indonesia Corp provided full support during the certification processes, which included preparation, documentation, training, and live test investigation.

The board of management involved in this process consisted of the Director of IT Solutions and Service Portfolio (CIO), AVP (assistant vice president) of Risk Management and Business Effectiveness, and AVP of Quality Management. The certification preparation journey took a long time. It started with the coordination meeting in London, UK, to plan the stages for ISO 27001 implementation. The process of document preparation and SoA establishment was done by Telkom's Infratel Division and Access Division.

The hand-over of the ISO 27001 certificate from TUV Rheinland Indonesia as the certification body to Telkom was conducted at Grha Caraka Citra Suite, where the certificate was handed over by the General Manager of TUV Rheinland Indonesia, Caterina Castellanos to the Director (CIO) of Telkom Indonesia Corp.

*"We hope that this certificate can be a milestone in winning the market as was the mandate and commitment from the meeting of the board of management of Telkom Group at 2011"* Telkom CIO said (bumn.go.id, 2012).

### 6.4.3.4    Marketing Aspects

The main aim of information security adoption is to protect business assets and to support the achievement of business goals (Dillon, 2007). Sherwood et al. (2005) adopted a multidimensional and enterprise-wide approach to information security and included in the scope of information security other aspects such as marketing and customer service. Information security literacy enables business by increasing its competitiveness, improves efficiency of business processes through the exploitation of new technologies, and increases trust from partners and customers.

Cherdantseva & Hilton (2013) declared that the protection of business assets and assistance with the achievement of business goals is the main aim of information security. Pipkin (2000) and Sherwood et al. (2005), stated that security adoption should be viewed as one of the organization's efforts that should involve a multi-discipline point of view: to improve the corporate selling point (Kotler, 2002), corporate imaging and branding (Dwyer, 1987), winning the competitive edge (Morrison et al., 2003), as a marketing tool, to increase corporate profitability, build customer trust, and make loyal customers (Brown et al., 2000). Moreover, Kotler (2002) stated, *"it is obvious that*

*business organization–producer, are inter dependencies to their loyal custom-*
*ers for the business sustainability"*. Customer loyalty is all about attracting the
right customers by winning their trust and making it convenient for them to
conduct business relationships with the producer (Combes & Patel, 1997).

The shifting role of information security from the technical to the broad
multidimensional discipline is also supported by Lacey (2011), who recounted
that information security *"draws on a range of different disciplines: computer
science, communications, criminology, law, marketing, mathematics and
more"*. He indicated the importance of technologies for protection of information, and emphasizes importance of the human factor which is based on the
fact that all technologies are designed, implemented and operated by people. In
addition to the human factor, Lacey also considered how organizational culture
and policies affect information security adoption.

*There is a strong correlation between the organization's information
security adoption and customer trust with the revenue and profitability of
the organization (Telkom Indonesia - 2012).*

*We will not be able to sell our services and products if we cannot
ensure that customer privacy is highly assured and safe. However, it is
necessary to obtain information security certification as a part of the
branding strategy that will convey a strong message that our company
is very secure to make customers comfortable in their investment, buying products and services from us (Bank Permata, 2012).*

Information security has spawned new potentials in business and re-
branding businesses, which could be implemented through technology
ability. However, re-branding is an economic-management and social
event as well as a strategy through which customers' demand and providers' supply must be balanced. Here, segmentation, targeting and positioning (STP) scenarios are needed for the corporation's new paradigm
approach through information security and trust issues. STP shows the
chronological dependency of the different activities. Segmenting is the
process of dividing the market into segments based on customer characteristics and needs of information security. Targeting is the analysis that leads
to a list of segments that is most attractive to target and have a good chance
of leading to a profitable market share. Positioning is the process by which
marketers try to create an image or identity in the minds of their target
market for its product or brand that complies with the security standard.

Telkom, as ISO 27001 holder, claims that as a direct influence of compliance it has increased corporate confidence, security guarantees, and customer trust. As a result, Telkom Indonesia has possibly increased their market size to approximately 50–60% market share in telecommunications (mobile phone) from approximately 220 million potential customers.

*Kompas, a well-known Indonesian newspaper, reported over the weekend that Indonesia's biggest telecommunication company, Telkom Indonesia, is looking to expand its business to seven countries in the future through its subsidiary companies. This is part of Telkom's vision to become a global player. At the moment, besides Indonesia as its main market, Telkom also operates in Singapore, Hong Kong, and Timor Leste (also known as East Timor). The company is conducting assessments in Australia and Myanmar for potential expansion programs. Telkom's cellular product, Telkomsel, was officially launched in Timor Leste on January 17th. Telkomsel has prepared a $50 million investment to be used until 2015 in order to build 3G infrastructures in Timor Leste. Nowadays with ISO 27001 compliance, Telkom has become a company whose security is recognized by their customers and this will probably enlarge their market share (Enricko -Tech in Asia, 2013).*

### 6.4.4   PART IV: SP/SQ MEASUREMENT

#### 6.4.4.1   Software Overview

ISM's development process is based on URS[14] as a document that specifies requirements to construct its features and modules. The details of ISM features to support e-assessment and e-monitoring as requested by users are as shown in Table 6.6.

#### 6.4.4.2   Cause-Effect Analysis

A cause and effect analysis is a method of analyzing an event or a problem devised. The chart used to represent the information gathered in such an analysis is also known as an Ishikawa diagram. It represents the various causes related to an event in the form of arms leading toward an event.

[14] User requirement(s) specification is a document usually used in software engineering that specifies the requirements the user expects from software to be constructed in a software project.

**TABLE 6.6**   Features and Functional Requirement

| Features | Function |
|---|---|
| RISC investigation tools | The software should be able to conduct evaluation of controls, clauses, domains, and top domain level. |
| Strength-Weakness Analysis | This feature is to determine the implemented controls and analyze controls that have not been implemented. |
| Chart and Histogram | The features function as an indicator of RISC investigation achievement level, compared with the ideal conditions that are required for compliance with ISO 27001. |
| Gaps Analysis | Gaps analysis is to show the distances between the current RISC achievement and ideal conditions of ISO 27001 compliance. |
| 21-controls comparison | This feature is to describe the achievements of RISC investigation, which consists of 6 domains, 7 clauses, 21 essential controls, 144 assessment issues and 256 refined questions. |
| Firewall | Firewall feature controls the incoming and outgoing network traffic by analyzing the data packets and determining whether they should be allowed through or not. A firewall feature establishes a barrier between a trusted, secure internal network and another network. |
| Process Information | This feature functions to further detect malicious software (malware) that is detected by the firewall. The feature "process information" allows users learn more about the potential suspects. In this module each suspect will be seen in more detail, such as; process ID, file name accessed that is running, set size, threads, parent process, and file type. |
| Network Detection | This feature functions as workstation monitoring that connects to the entire network, wired or wireless network. In case there are unrecognized clients joining to the network, then the admin can perpetrate preventive action to protect the related network, data, and information associated. |
| Port Detection | Port detection is an application-specific or process-specific software construct serving as a communications endpoint. Its purpose is to uniquely identify different applications or processes running on a single computer and thereby enable them to share a single physical connection to a packet-switched network like the Internet. Ports are also used by hackers as a way in to do the demolition on network components, information, and data. Therefore monitoring of absolute ports is required, particularly to prevent the things that are not desired as stated. |

In the case of ISM's development, cause-effect approach is used to describe the problems, causes, and solutions, as an adaptation of user expectations. There are six main problems addressed by ISM's development (Figure 6.29 and Table 6.7): (1) Lack of a formal method for RISC measurement. (2) Lack of software application for assisting organizations for RISC investigation. (3) The need for GUI-based software development for easy operation. (4) The users difficulty to conduct information security self-investigation within each level of assessment issues, controls, clauses, and domains. (5) The organizations' difficulty to focus on the weak parts of the achievement. (6) The Users' need for a monitoring system to check real-time security status against suspected breaches.

### 6.4.4.3   The Eight Fundamental Parameters Measurement

The Eight fundamental parameters (8FPs[15]) consist of software performance and quality indicators. The 8FPs reflect ISM's performance, behavior, function, and back-office programming to handle the workload of self assessment and monitoring. Those indicators are shown in Table 6.8. In this study, users claim that the level of reliability and usability of ISM to understand ISO 27001 was 67%, highly recommended. ISM's function for RISC self-assessment is shown at 90% recommended. This result shows that ISM can accelerate the organization's ability to obtain ISO 27001 certification.

Table 6.8 and Figure 6.30 shows the eight parameters. The average score from cluster-I (C-I), cluster-II (C-II), and cluster-III (C-III) are mentioned. The users' measurement results regarding ISM's functions, features and performance are provided. Illustrative measurements are shown in the following figures. Figure 6.31 shows the state of the eight parameters and the moving average. The overall score of all parameters is shown at "2.7" points. The parameters "ISM function as information security self-assessment", "ISM for understanding terms and concepts", "ISM graphical user interface and user friendliness", "Analysis Precision of ISM", "Final Precision of ISM" and "Performance of ISM" scored highest at "3", and the parameter of "the benefit of ISM to help organizations understand ISO

---

[15] All 8FPs reflected of ISM performance, behavior, function, and back-office programming to handle certain workload of self assessment and monitoring.

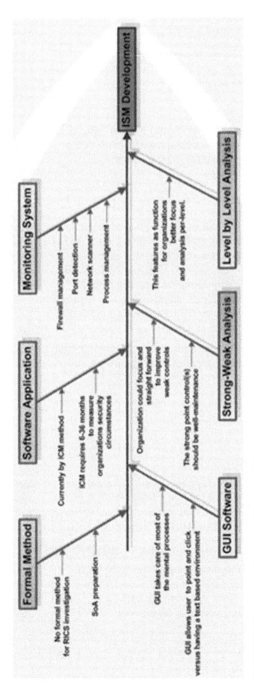

**FIGURE 6.29**  Analysis and solution for ISM.

**TABLE 6.7**   Analysis and Solution for ISM

| Problem | Causes and Effects | Solution |
|---|---|---|
| The lack of a formal method to measure RISC level. | Organizations are highly dependent on their consultants to accomplish all stages of compliance | To propose an applied framework that provides a formal method to do RISC investigation |
| The lack of software applications that can assists organizations for RISC investigation. | There is an existing method called ICM, but this method requires much time to accomplish security compliance | The need to develop semi-automated software to conduct RISC investigation |
| The need to embed GUI programming approach. | GUI allows user to point and click versus having a text based environment. As users do not have to be trained extensively to use software with user friendly GUI environment, it saves training time. For individuals, the advantages are also great because GUI takes care of most of the mental processes. | Develop a software with visual programming language |
| The users find it difficult to investigate security circumstances in each level of assessment issues, controls, clause, and domain. | This functions so that organizations can better focus and analyze per-level. | Developed a module that can describe readiness level layer-by-layer. |
| The organizations find it difficult to focus on the weak parts of the achievement. | Nowadays, the organizations are using the ICM tool, the consequence is the large amount of time required to review all controls to find their weak points. | Develop features that can directly show the weak points of an organization towards ISO 27001 implementation. |
| The need for a monitoring system to detect suspected information security breaches. | The monitoring system is the implementation of technical controls of ISO 27001 as a real-time traffic and software monitoring of the network. | Develop a monitoring system that embeds with RISC investigation. |

**TABLE 6.8**  ISF and ISM as Corporate Feedback and Assessment

| | 0 | 1 | 2 | 3 | 4 |
|---|---|---|---|---|---|
| ISM functions as information security self-assessment | | | *C-I: 2*<br>*C-II: 2.2*<br>*C-III: 2*<br><br>*Average:*<br>*2.1* | | |
| The benefit of ISM to help organizations understand ISO 27001 controls | | | *C-I: 2*<br>*C-III: 2*<br><br>*Average: 2.1* | *C-II: 2.8* | |
| ISM can be used to understand terms and concepts | | | *C-III: 2.5*<br><br>*Average: 2.9* | *C-I: 3*<br>*C-II: 3.2* | |
| ISM features | | | *C-I: 2*<br><br>*Average: 2.7* | *C-II: 3*<br>*C-III: 3* | |
| ISM graphical user interface and user friendliness | | | *C-III: 2.5*<br><br>*Average: 2.9* | *C-I: 3*<br>*C-II: 3.2* | |
| Analysis precision produced by ISM | | | *C-I: 2*<br>*C-III: 2.5*<br><br>*Average: 2.5* | *C-II: 3* | |
| Final result precision produced by ISF-ISM | | | *C-I: 3*<br>*C-III: 2.5*<br><br>*Average: 2.8* | *C-II: 2.8* | |
| ISM Performance overall | | | | *C-I: 3*<br>*C-II: 3*<br>*C-III: 3.5*<br>*Average:*<br>*3.2* | |
| *Average* | | | *2.7* | | |
| *C-I average* | | | *2.5* | | |
| *C-II Average* | | | *2.9* | | |
| *C-III Average* | | | *2.6* | | |

C-I = ISO 27001 holder; C-II = ISO 27001 ready; C-III = ISO 27001 consultant.

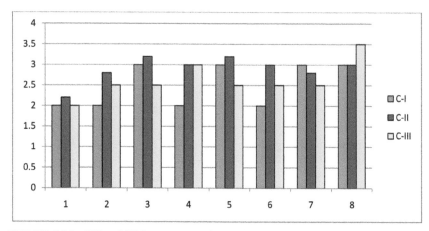

**FIGURE 6.30**   ISF and ISM as corporate feedback and assessment.

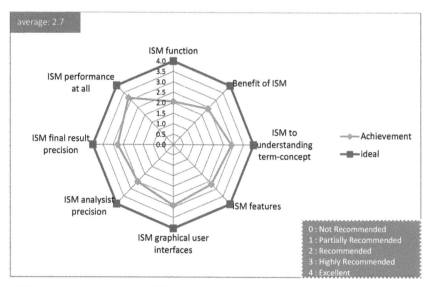

**FIGURE 6.31**   The result of SP/SQ measurement against 8FPs.

27001 controls" and "ISM features" scored lowest at "2". The basic eight parameters of the success indicators of ISM are given in the following section.

## 6.4.4.4  Security Culture

A security awareness policy is often met with resistance from end-users during implementation. The awareness program aims to change the organization's behavior and culture. Implementing an awareness program is the cheapest method when compared to security breach countermeasures. In order to be effective, security awareness should start from the board of management as a role model for information security literacy. This was shown in the case of Telkom Indonesia Corp during the compliance process that was fully supported by the board of management.

To create an effective security awareness program, *firstly* it is important to assess the strengths and weaknesses of the personalities of organization members, to easier identify which parts of the program require additional effort, resources or knowledge. *Secondly*, efficient communication and harmony between business processes and the information security program is needed to ensure the program is successful and durable. *Thirdly*, there is the need to consider the policies and standards in the organization.

Enforcement is the key for successful implementation. Here, ISM can help solve the hurdles of the awareness program. With its methodology and use of emerging technology, ISM has become a media for knowledge sharing and self-assessing for organizations to further analyze their information security.

Organizations, through the ICT department or the information security officer, should educate the employees in dealing with information to avoid breaches by unauthorized users. The organization can also install software to protect the information from any breaches. In this case, the e-monitoring system which is part of ISM has a role to resolve the issue to detect any suspected information security breaches.

### 6.4.4.5  Efficiency and Effectiveness

ISM is a semi-automated tool developed from ISF for organizations to check the closeness of their compliance with ISO 27001. Respondents agreed that ISM can guide and help them comprehend ISO 27001 terms and concepts, by mapping compliance problems such as essential controls, clauses, assessment issues in an organized way and it provides a practical approach to measure the readiness level to pursue ISO 27001 compliance. ISM helps respondents figure out the ISO 27001 structure easily. Furthermore it reduces ISO 27001 learning time and they can quickly assess the status of their information security issues. ISM definitely helps them learn critical status points, especially the weak and strong points of the current system through the achievement priority gap analysis feature, and histogram results feature and these features can provide an idea of compliance with ISO 27001.

Figure 6.32 shows ISM's influence on an organization's understanding of ISO 27001. There are three main benefits of ISM that improves the effectiveness of preparation stages to satisfy the standard, (1) ISM makes it easier for organizations to review their position and circumstances of their information security, (2) ISM reduces time, and (3) reduces barriers in the interpretation of technical terms and definitions.

The factor of time efficiency is indicated by the time spent on self-assessment of RISC investigation. 77% of users needed approximately 60 min, and 33% users required between 61–720 min to conduct RISC investigation. This fact is an indicator of efficiency through the improved understanding of information security business processes during RISC measurement.

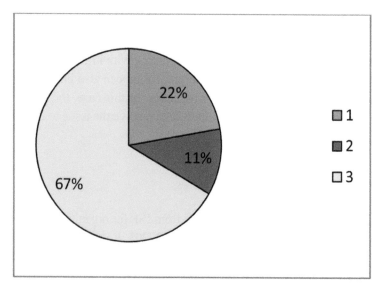

**FIGURE 6.32**   ISM influences

On the other hand, compared with the process that was conducted by Bank Indonesia, Telkom Indonesia during testing in 2012, and Kosutic (2013), RISC investigation and determining the potential strengths and weaknesses of the organizations' essential controls through ICM required 6–36 months.

### 6.4.4.6 Enhance Security Literacy

ISM makes it possible for organizations to assess and forecast their RISC level actively to find out their information security circumstances. The organization's information security self-awareness is an indicator for increasing its capabilities. All respondent clusters described that the information security self-awareness aspect provided by the ISM is very useful for further compliance processes.

Cluster-I (C-I) stated that ISM is a tool that can assist them to perform self-assessment, monitor and maintain their compliance (Figure 6.33, green line). Note that an ISO 27001 certification body will re-evaluate their compliance every two years. Consequently, a tool that can quickly provide compliance information to ISO 27001 controls, clauses (objectives) and domains is very useful. As software, ISM also monitors current circumstances associated with information security issues, including controls associated with the infrastructure management, recovery system, worst case scenarios, and risk management. In fact, ISM provides strong-weak points analysis and recommendations of priorities for systematic maintenance of ISO 27001 compliance. This means that the cost of maintenance will be much cheaper as the need and interaction with information security consultants can be minimized.

Cluster-II (companies that are currently pursuing ISO 27001 certification) stated that the ISM is helpful, particularly during periods of preparation, understanding terms, and SoA documentation for certification. In addition, pre-audit and pre-assessment can be performed to find out the gap between existing and expected condition required by the standard (Figure 6.33, blue line). ISM provides a semi-automatic assesses readiness level for compliance, before inviting assessors for ISO 27001 certification. With the approach offered, they could easily ensure the readiness in satisfying ISO 27001 controls rather than inviting evaluators to find out the gaps that may occur.

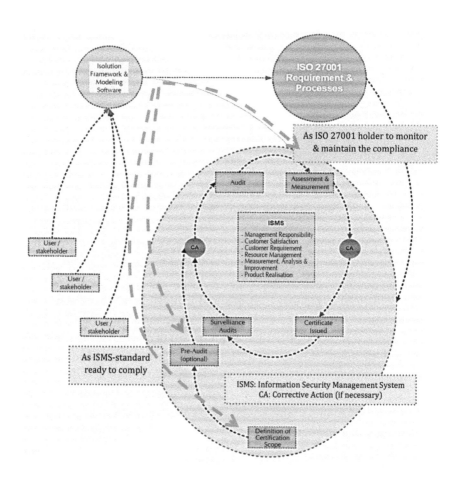

**FIGURE 6.33**   ISM as corporate self-assessment tool.

The respondents from cluster-III are the ICT information security consultants that are currently using an ICM form (either hard copy or soft copy format) which consists of security checklists to perform security assessment. It is the method to pursue with the parameters of the standard, following five steps of compliance; *risk assessment methodology* to define rules on how to perform compliance measurement; *risk assessment implementation* **to** find out which potential problems could happen, list all assets, then threats and vulnerabilities related to those assets, assess the impact and likelihood for each combination of assets/threats/vulnerabilities and finally calculate the level of risk; *treatment implementation* to find a solution for implementation; *assessment report* to document certain processes that have happened or were performed, for the purpose of audit and review; *risk treatment plan* to define exactly who is going to implement each control, including the timeframe and budget (Figure 6.34).

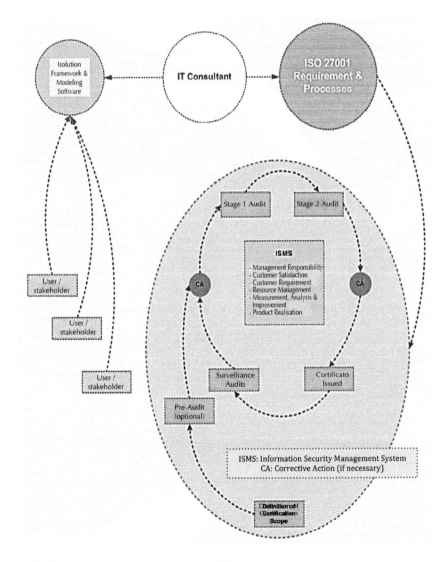

**FIGURE 6.34**    ISM as corporate tool for ICT consultant.

The checklist form is unable to explain the compliance level quickly. ISM will help consultants to provide systematic advice to those companies interested in pursuing an ISO 27001 certification processes.

## KEYWORDS

- five principles of security
- readiness and information security capabilities
- SP/SQ measurement
- software development methodologies

## CHAPTER 7

# CONCLUSIONS AND RECOMMENDATIONS

## CONTENTS

In a competitive business environment, customers' trust is one of the key success factors for business organizations to be competitive and grow. Customers' trust is highly related to the security of their information. Information security enables business by increasing its competitiveness. The main aim of information security is to protect business assets and to support the achievement of business goals. Security of information provides advantages such as improved efficiency of business processes due to the exploitation of new technologies and increased trust from partners and customers. As information security is very important, it has to be managed properly through an information security management system. With an information security management system, a business organization can demonstrate to partners and customers that it has identified and measured potential security risks and implemented a security policy and controls that will mitigate these risks.

Many organizations face internal and external challenges to implement an information security management system to protect their information assets. Some notable barriers and challenges are the lack of available human resources to support the implementation, the technical issues of information security, and selecting a suitable standard to comply with. These challenges are not easy to overcome unless those organizations address all information security issues using an appropriate framework to face them systematically.

The main goal of this study is to assist organizations face these challenges by constructing a novel framework that can be used to develop a system to measure the closeness of their information security status with an Information Security Management System (ISMS) standard (a compliance level) and subsequently guide them on how to address information security issues effectively. As such, this study covered three important aspects, namely (1) a comparative study among ISMS standards to describe their position and roles, (2) a framework for mapping of controls, clauses and assessment issues with a chosen ISMS standard (ISO 27001), (3) developing an application software to investigate the RISC level of security adoption.

## 7.1   THE MAJOR FINDINGS

It is important to note that securing and keeping information from parties who do not have authorization to access such information is an extremely important issue. To address this issue, it is essential for an organization to implement an ISMS standard such as ISO 27001 to address the issue comprehensively. The authors have constructed a novel security framework (ISF) and subsequently used this framework to develop software called Integrated Solution Modeling (ISM), a semi-automated system that will greatly help organizations comply with ISO 27001 faster and cheaper than other existing methods. In addition, ISM does not only help organizations to assess their information security compliance with ISO 27001, but it can also be used as a monitoring tool, helping organizations monitor the security statuses of their information resources as well as monitor potential threats. ISM is developed to provide solutions to solve obstacles, difficulties, and expected challenges associated with literacy and governance of ISO 27001. It also functions to assess the RISC level of organizations towards compliance with ISO 27001.

The study has successfully developed a framework that embeds information security self-awareness and literacy to extend the role of organizations in their own steps toward compliance. The framework offers a model for an integrated solution for security literacy, self-assessment, and real time monitoring. This solution empowers an organization by increasing its understanding of information security and the ISO 27001 standard, reducing the need for consultants and improving confidence when facing the scrutiny of the standards body. ISM is different from the existing methods for helping organizations comply with ISO 27001 in five ways. First, the software is multiuser and it is developed from the framework (ISF). The ISM architecture, design, and configuration are fine-tuned to the client-server system. Client-server applications are a rapidly growing trend in the interconnected environment, especially in an organization with national and global capacities. As such, ISM is expected to be the center point for the RISC investigation and network real-time monitoring.

Second, ISF has shown positive effects on two clusters of companies involved in the research: the ISO 27001 holders' cluster and the ISO 27001 ready cluster. A notable change is the active participation during the ISO 27001 compliance process. With the existing methods, organizations participate passively in the ISO 27001 compliance process by following what consultants advise during adoption stages. With the new method (ISM), organizations actively participate during the compliance process. This has led to the improvement of their information security ability, capability, and literacy.

Third, for ISO 27001 holders, ISM is a tool that could help them to perform self-assessment, monitor and maintain their compliance. Note that an ISO 27001 certification is normally re-evaluated every two years. ISM can quickly provide compliance information with regards to ISO 27001 controls, clauses (objectives), and domains, which need to be reviewed during the process of renewal of their compliance. ISM also monitors current circumstances associated with information security issues, including the infrastructure management, recovery system, worst case scenarios, and risk management. In fact, ISM provides the strong-weak points analysis and offer recommendations of priorities for systematic maintenance of ISO 27001 compliance. Consequently, the cost for maintenance and renewal will be much cheaper than existing methods as the need for security consultants is minimized.

Fourth, for the ISO 27001-ready organizations, ISM is very helpful, particularly during the preparation stage as it helps improve the understanding of ISO 27001 terms as well as helping to construct the SoA document for the certification process. In addition, both pre-audit and pre-assessment can be performed using ISM to find out the gaps between existing and expected conditions required by ISO 27001. ISM provides a semi-automatic assessment for level of compliance (readiness level), before inviting assessors for an ISO 27001 certification. This approach makes it easier for organizations to measure their RISC level to satisfy ISO 27001 controls themselves rather than inviting a consultant to find out the gaps.

Fifth, for the ISO 27001 consultants, the current practice is to use a checklist form (either hard copy or soft copy format) (ICM) consisting of security checklists to perform a security assessment. It is a manual method to pursue ISO 27001 compliance through parameter checking following five steps of compliance: *risk assessment methodology* to define rules on how to perform compliance measurement; *risk assessment implementation* to finding out which potential problems could happen, list all assets, then threats and vulnerabilities related to those assets, assess the impact and likelihood for each combination of assets/threats/vulnerabilities and finally calculate the level of risk; *treatment implementation* to find a solution to implement; *assessment report* to document controls that have been performed for the purpose of audit and review; and *risk treatment plan* to define exactly who is going to implement each control, including the timeframe and budget.

To find out the respondents' tendencies toward the framework and its application, survey instruments were specifically developed and administrated to a sample of participants. The data gathered from the survey were analyzed to provide ideas on how e-assessment and e-monitoring can help organizations assess and monitor their security compliance with ISO 27001.

The results of the survey were also used to improve the framework and further ISM development that would incorporate the expectations of the users as a RISC investigation tool. The users tried and tested ISM for approximately 12 months. During this period the impact of security literacy and self-awareness and also ISM's performance were monitored and measured.

The performance, quality, features, reliability, and usability of ISM have been highly valued by users during the trial and test. The overall score of all parameters is 2.70, which means that users recommend (close to "highly recommended") the usage of ISM for their organizations. This means that our framework (ISF) is effective and certainly its implementation is useful for organizations to assess RISC toward ISO 27001 compliance.

## 7.2 RECOMMENDATIONS AND FUTURE RESEARCH DIRECTIONS

This study incorporates two main modules of RISC investigation: e-assessment and e-monitoring. E-assessment is to investigate the organization's closeness to the ideals of the security standard by measuring ISO 27001's security controls. On the other hand, the e-monitoring module provides real-time monitoring for suspected information security breaches. These modules were integrated comprehensively to find out the level of organizations' security adoption. ISM differentiates the concept of Security Assessment Management (SAM) and Security Monitoring Management (SMM). These concepts refer to the users' assessment towards essential controls of ISO 27001 (in e-assessment) and real-time monitoring for suspect security breaches (in e-monitoring). With the promising features and user acceptance of its performance, quality, reliability and usability, it is highly recommended for organizations to consider adopting ISM as it has been valued as an effective, efficient tool to assist organizations to comply with ISO 27001 and helps to set a clear strategy to get an ISO 27001 certificate confidently.

The study triggers a future research direction in security strategy which would focus on enhancing organization security, change management associated with the implementation of security cultures, incident management dealing with the impact and escalation of disaster, software release management, IT service continuity management and configuration management to handle security incidents.

The future direction of this study can also accommodate and customize ISF to fit with other standards such as BS 7799, COBIT, ITIL, and others.

ISF was designed as an adaptable framework that can be implemented and extended to other standards by customizing its domains. ISF could possibly be implemented to others by following mapping stages through the grouping of controls to respective domains in each of these standards. To conduct assessment, the ISF mathematical formula should be adapted to measure against standards other than ISO 27001.

During the testing stage and interaction, respondents voiced ideas further enhance the functionality of ISM in future developments. Users suggested the integration of more complete and robust security support features, particularly the integration of an information security decision support system, an expert system and a security pattern recognition system, complemented with a knowledge inference and learning system to emulate the decision making ability of a human expert. This software ability could solve compliance barriers, create an early warning system for suspected security breaches and help enhance strategic planning of information security.

The ISM prototype was implemented using client-server programming technology. It will be interesting to extend the coverage to support mobile application emerging technologies (apps) that adopts the concept of mobile RISC measurement anytime and anywhere.

There are limitations from the study that need to be addressed in the future. First, the objects of research were organizations bound with the predeterminant security cultures and awareness that may make the finding differ from others places. As such, it is necessary to test the ISM in a bigger sample in several countries.

## 7.3   CONCLUDING REMARKS

Securing information resources from unauthorized access is important to maintain business operations in today's world of information technology, whether it is to protect assets or used as a marketing tool. Information security needs to be managed in a proper and systematic manner as it is quite complex. One of the effective ways to manage information security is to comply with an information security management standard. There are many standards recognized for information security such as ISO 27001,

BS 7799, ITIL, COBIT, and PCIDSS, the most widely recognized and used being ISO 27001. Therefore, it is important for an organization to implement ISO 27001 to address information security issues comprehensively. Unfortunately, there are barriers to compliance with the ISO 27001, since the compliance methods are complex, time consuming and expensive. A new method, preferably supported by a semi-automated tool, is much welcome.

ISF is a framework that addresses the need for a method to assess an organization's readiness level and compliance with an information security standard (ISO 27001). ISF was developed to tackle issues which are not properly addressed by the existing security frameworks (9STAF) such as measurable quantitative valuation, mathematical modeling, algorithm, and applicability for software development. ISF is designed to help organizations perform ISO 27001 compliance projects efficiently. Based on ISF, a semi-automated tool was developed to assess the readiness of an organization to comply with ISO 27001 and subsequently use the tool to assess the potential threats, strengths and weaknesses for efficient and effective implementation of ISO 27001.

As an implementation of ISF, ISM uses a bottom-up approach to measure ISO 27001 essential controls. Those controls are recursively counted from the lowest level to the top level of domains. ISF has a unique design approach that can be implemented in the measurement of compliance with other standards. Qualifying for an ISO 27001 certificate is one of an organization's efforts to protect its information resources and at the same time boost its selling point to its customers by winning their trust. ISM has made the assessment towards ISO 27001 compliance more practical, easier and systematic. With the existing methods, organizations fully depend on third parties or consultants for the compliance process. However, with ISM, the dependency on consultants is reduced. Hence, organizations will not only save money, but also gain more understanding on the compliance process.

To find out the effectiveness of ISM, a comprehensive testing and evaluation were conducted. The result is very promising as ISM was well evaluated by the respondents and accepted as a useful tool to help companies systematically plan to acquire ISO 27001 certifications. The user response towards the performance, quality, features, reliability, and usability has been highly positive.

We have found that in the cases of our respondents, ISM could conduct RISC investigation within 12 h, which is much better than the ICM approach that required approximately 12 months for the investigation. This means our framework is effective, efficient, significant, and certainly its implementation is useful for organizations to assess their compliance with ISO 27001. Hence, they can then set a clear strategy to obtain ISO 27001 certification with confidence.

## KEYWORDS

- integrated solution modeling
- risk assessment implementation
- risk assessment methodology
- risk treatment plan
- security assessment management
- security monitoring management

# BIBLIOGRAPHY

Abu Saad, B., Saeed, F. A., Alghathbar, K., & Khan, B. (2011). Implementation of ISO 27001 in Saudi Arabia–Obstacles, motivations, outcomes, and lessons learned.

Aceituno, V. (2005). On Information Security Paradigms. *ISSA Journal, September, 24*, 225–229.

Al Omari, L., Barnes, P. H., & Pitman, G. (2012, December). Optimising COBIT 5 for IT governance: examples from the public sector. In: *Proceedings of the ATISR 2012: 2nd International Conference on Applied and Theoretical Information Systems Research (2nd. ATISR2012)*. Academy of Taiwan Information Systems Research.

Al Osaimi, K., Alheraish, A., & Bakry, S. H. (2008). STOPE–based approach for e-readiness assessment case studies. *International Journal of Network Management, 18*(1), 65–75.

Alfantookh, A. (2009). An Approach for the Assessment of The Application of ISO 27001 Essential Information Security Controls. *Computer Sciences, King Saud University*.

Aljabre, A. (2012). Cloud computing for increased business value. *International Journal of Business and Social Science, 3*(1), 234–239.

Almunawar, M. N., Anshari, M., & Susanto, H. (2013). Crafting strategies for sustainability: how travel agents should react in facing a disintermediation. *Operational Research, 13*(3), 317–342.

Alshitri, K. I., & Abanumy, A. N. (2014, May). Exploring the Reasons behind the Low ISO 27001 Adoption in Public Organizations in Saudi Arabia. In: *2014 International Conference on Information Science and Applications (ICISA)*, IEEE. pp. 1–4.

Amoroso, E. G. (1994). *Fundamentals of Computer Security Technology*. Prentice-Hall, Inc.

Anderson, B., & Adey, P. (2011). Affect and Security: Exercising Emergency in UK Civil Contingencies'. *Environment and Planning D: Society and Space, 29*, 1092–1109.

Anderson, E., & Weitz, B. (1989). Determinants of continuity in conventional industrial channel dyads. *Marketing Science, 8*(4), 310–323.

Anderson, J. M. (2003). Why we need a new definition of information security. *Computers & Security, 22*(4), 308–313.

Anderson, K. (2006). IT Security Professionals Must Evolve for Changing Market. *SC Magazine*, 12, 2006.

Anderson, R. (2001). Why information security is hard-an economic perspective. In: *Proceedings 17th Annual Computer Security Applications Conference, 2001. ACSAC 2001*. IEEE. pp. 358–365.

Anderson, R., & Moore, T. (2006). The Economics of Information Security. *Science, 314*(5799), 610–613.

Anttila, J., Kajava, J., & Varonen, R. (2004). Balanced integration of information security into business management. In: *Proceedings. 30th Euromicro Conference, 2004*. IEEE. 558–564.

Avizienis, A., Laprie, J. C., Randell, B., & Landwehr, C. (2004). Basic concepts and tax-onomy of dependable and secure computing. *IEEE Transactions on Dependable and Secure Computing, 1*(1), 11–33.

Bailey, M., Oberheide, J., Andersen, J., Mao, Z. M., Jahanian, F., & Nazario, J. (2007). Automated classification and analysis of internet malware. In: *Recent Advances in Intrusion Detection*. Springer Berlin Heidelberg. pp. 178–197.

Baker, L. B., & Finkle, J. (2011). Sony PlayStation suffers massive data breach. *Reuters, April, 26.*

Baker, M., Walker, O., Mullins, J., Boyd, H., Larreche, J., & Cravens, D. (1996). Market-ing strategy. *International Encyclopedia of Business and Management*, 3333–3347.

Bakers, S. (2011). Facebook statistics by country. *URL: http://www.socialbakers.com/facebook-statistics/-abgerufenam, 17*, 2011.

Bakers, S. (2013). *Social Bakers*. Retrieved March, 14, 2013.

Bakos, J. Y. (1991). A strategic analysis of electronic marketplaces. *MIS quarterly*, 295–310.

Bakos, J. Y. (1997). Reducing buyer search costs: Implications for electronic marketplaces. *Management Science, 43*(12), 1676–1692.

Bakry, S. H. (2003a). Toward the development of a standard e-readiness assessment policy. *International Journal of Network Management, 13*(2), 129–137.

Bakry, S. H. (2003b). Development of security policies for private networks. *International Journal of Network Management, 13*(3), 203–210.

Bakry, S. H. (2004). *Development of E-Government: A STOPE View. International Journal of Network Management, 14*(5), 339–350.

Bakry, S. H., & Bakry, F. H. (2001). A strategic view for the development of e-business. *International Journal of Network Management, 11*(2), 103–112.

Baraghani, S. N. (2008). Factors influencing the adoption of internet banking. *Lulea Uni-versity of Technology.*

Benjamin, R., & Wigand, R. (1995). Electronic markets and virtual value chains on the information superhighway. *Sloan Management Review (Winter, 1995).*

Besnard, D., & Arief, B. (2003). *Computer Security Impaired by Legal Users*. University of Newcastle upon Tyne, Computing Science.

Bitazar, A. (2009). About ISO 27001 Benefits and Features. Obtained from http://www.articlesbase.com.

Blakley, B., McDermott, E., & Geer, D. (2001). Information security is information risk management. In: *Proceedings of the 2001 Workshop on New Security Paradigms*. ACM. 97–104.

Blessing, L. T. M., Chakrabarti, A., & Wallace, K. M. (1998). An overview of descriptive studies in relation to a general design research methodology. In: *Designers*. Springer London. pp. 42–56.

Blyth, A., & Kovacich, G. L. (2001). *What is Information Assurance?*. Springer London. pp. 3–16.

Boehm, B. W., Brown, J. R., & Lipow, M. (1976). Quantitative evaluation of software quality. In: *Proceedings of the 2nd International Conference on Software Engineer-ing*. IEEE Computer Society Press. pp. 592–605.

Boehm, B., & Hansen, W. (2001). The spiral model as a tool for evolutionary acquisition. *CrossTalk, 14*(5), 4–11.

Boehm, B., & Hansen, W. J. (2000). *Spiral Development: Experience, Principles, and Refinements* (No. CMU/SEI-2000-SR-008). Carnegie-Mellon University Pittsburgh, PA, Software Engineering Inst.

Boehmer, W. (2008). Appraisal of the Effectiveness and Efficiency of an Information Security Management System Based on ISO 27001. *SECURWARE, 8*, 224–231.

Bonner, E., O'Raw, J., & Curran, K. (2013). Implementing the Payment Card Industry (PCI) Data Security Standard (DSS). *TELKOMNIKA (Telecommunication Computing Electronics and Control), 9*(2), 365–376.

Boss, S. R., Kirsch, L. J., Angermeier, I., Shingler, R. A., & Boss, R. W. (2009). If someone is watching, I'll do what I'm asked: mandatoriness, control, and information security. *European Journal of Information Systems, 18*(2), 151–164.

Bowen, T. P., Wigle, G. B., & Tsai, J. T. (1985). *Specification of software quality attributes.* Rome Air Development Center, Air Force Systems Command.

Boyce, J. G., & Jennings, D. W. (2002). *Information Assurance: Managing Organizational IT Security Risks.* Butterworth-Heinemann.

Brand, K., & Boonen, H. (2007). *IT Governance Based on Cobit 4. 1: A Management Guide.* Van Haren Publishing.

British Standard (BS). (2012). An Overview of British Standard: BS 7799. Obtained from http://www.bsigroup.com/en/Standardsand-Publications/About-BSI-British-Standards/.

Brown, J. R., Dev, C. S., & Lee, D. J. (2000). Managing marketing channel opportunism: the efficacy of alternative governance mechanisms. *Journal of Marketing, 64*(2), 51–65.

Bulgurcu, B., Cavusoglu, H., & Benbasat, I. (2009). Roles of information security awareness and perceived fairness in information security policy compliance.

Bulgurcu, B., Cavusoglu, H., & Benbasat, I. (2010). Information security policy compliance: an empirical study of rationality-based beliefs and information security awareness. *MIS Quarterly, 34*(3), 523–548.

BUMN.go.id. (2012). The Adoption of ISO 27001 of Telkom Indonesia: A Key Performance for Customer Trust. Obtained from: www.bumn.go.id.

Calder, A., & Watkins, S. (2012). *IT Governance: An International Guide to Data Security and ISO 27001/ISO27002.* Kogan Page Publishers.

Calder, A., & Watkins, S. G. (2010). *Information Security Risk Management for ISO 27001/ISO27002.* It Governance Ltd.

Cavano, J. P., & McCall, J. A. (1978). A framework for the measurement of software quality. *ACM SIGSOFT Software Engineering Notes, 3*(5), 133–139.

Cavusoglu, H., Cavusoglu, H., & Raghunathan, S. (2004a). Economics of ITSecurity Management: Four Improvements to Current Security Practices. *The Communications of the Association for Information Systems, 14*(1), 37.

Cavusoglu, H., Mishra, B., & Raghunathan, S. (2004b). A model for evaluating IT security investments. *Communications of the ACM, 47*(7), 87–92.

Cavusoglu, H., Mishra, B., & Raghunathan, S. (2004c). The effect of internet security breach announcements on market value: Capital market reactions for breached firms and internet security developers. *International Journal of Electronic Commerce, 9*(1), 70–104.

Cherdantseva, Y., & Hilton, J. (2013). Information Security and Information Assurance. The Discussion about the Meaning, Scope and Goals. *Organizational, Legal, and Technological Dimensions of IS Administrator.* IGI Global Publishing.

Cherdantseva, Y., Rana, O., & Hilton, J. (2011). Security Architecture in a Collaborative De-Perimeterised Environment: Factors of Success. *ISSE Securing Electronic Business Processes*, 201–213.

Clark, D. D., & Wilson, D. R. (1987). A Comparison of Commercial and Military Computer Security Policies.

Cockburn, A., & Highsmith, J. (2001). Agile software development: The people factor. *Computer*, *34*(11), 131–133.

Combes, G. C., & Patel, J. J. (1997). Creating lifelong customer relationships: Why the race for customer acquisition on the Internet is so strategically important. *iword*, *2*(4), 132–140.

Committee on National Security Systems, CNSS. (2004). 4009, "National Information Assurance Glossary," Committee on National Security Systems, May 2003. *Formerly NSTISSI, 4009*.

Cooper, H. (1998). Synthesizing Research: A Guide for Literature Reviews Common Criteria (CC) for Information Technology Security Evaluation. Part 1: Introduction and general model. Version 3.1. Revision 3. July 2009.

Costa, L., & D'Amico, R. (2011). Malware Detection and Prevention Platform: Telecom Italia Case Study. In: *ISSE 2010 Securing Electronic Business Processes*, pp. 203–213. Vieweg + Teubner.

Cusumano, M. A., & Smith, S. A. (1995). Beyond the Waterfall: Software Development at Microsoft.

D'Arcy, J., Hovav, A., & Galletta, D. (2009). User awareness of security countermeasures and its impact on information systems misuse: A deterrence approach. *Information Systems Research*, *20*(1), 79–98.

D'Arcy, S. P., & Brogan, J. C. (2001). Enterprise risk management. *Journal of Risk Management of Korea*, *12*(1), 207–228.

Da Veiga, A., & Eloff, J. H. (2010). A framework and assessment instrument for information security culture. *Computers & Security*, *29*(2), 196–207.

Dark, M. J., Ekstrom, J. J., & Lunt, B. M. (2005). Integration of information assurance and security into the IT2005 model curriculum. In: *Proceedings of the 6th Conference on Information Technology Education*. ACM, pp. 7–14.

De Haes, S., Van Grembergen, W., & Debreceny, R. S. (2013). COBIT 5 and enterprise governance of information technology: Building blocks and research opportunities. *Journal of Information Systems*, *27*(1), 307–324.

De Haes, S., Van Grembergen, W., & Debreceny, R. S. (2013). COBIT 5 and enterprise governance of information technology: Building blocks and research opportunities. *Journal of Information Systems*, *27*(1), 307–324.

Dellinger, A. (2005). Validity and the review of literature. *Research in the Schools*, *12*(2), pp. 41–54.

Dellinger, A. B., & Leech, N. L. (2007). Toward a unified validation framework in mixed methods research. *Journal of Mixed Methods Research*, *1*(4), 309–332.

Detik.com. (2013). Indonesia – Australia in Wiretapped issues (www.detik.com).

Dhillon, G. (2007). *Principles of Information Systems Security: Text and Cases*. New York: Wiley. 97–129.

Dick, A. S., & Basu, K. (1994). Customer loyalty: Toward an integrated conceptual framework. *Journal of the Academy of Marketing Science*, *22*(2), 99–113.

Dictionary, O. E. (2013). Oxford English Dictionary.

Diefenbach, T. (2009). Are case studies more than sophisticated storytelling?: Methodological problems of qualitative empirical research mainly based on semi-structured interviews. *Quality & Quantity, 43*(6), 875–894.

Dietrich, C. J., Rossow, C., & Pohlmann, N. (2013). CoCoSpot: Clustering and recognizing botnet command and control channels using traffic analysis. *Computer Networks, 57*(2), 475–486.

Dlamini, M. T., Eloff, J. H., & Eloff, M. M. (2009). Information security: The moving target. *Computers & Security, 28*(3), 189–198.

Doherty, N. F., & Fulford, H. (2006). Aligning the information security policy with the strategic information systems plan. *Computers & Security, 25*(1), 55–63.

Dubois, É., Heymans, P., Mayer, N., & Matulevičius, R. (2010). *A Systematic Approach to Define the Domain of Information System Security Risk Management.* In Intentional Perspectives on Information Systems Engineering (pp. 289–306). Springer Berlin Heidelberg.

Dwyer, F. R., Schurr, P. H., & Oh, S. (1987). Developing buyer-seller relationships. *The Journal of Marketing,* 11–27.

Easttom, W. C. (2012). *Computer Security Fundamentals.* Pearson Education India.

Eloff, J. H. P., & Eloff, M. M. (2005). Information security architecture. *Computer Fraud & Security, 2005*(11), 10–16.

Eloff, M. M., & Solms, S. H. (2000). Information security management: A hierarchical framework for various approaches. *Computers & Security, 19*(3) 243–256. doi: 10.1016/S0167-4048(00)88613-7. Elsevier.

FinanceToday.com. (2013). Indonesian's ICT consumer behavior. Obtained from www. financetoday.com.

Fink, D. (1994). A Security Framework for Information Systems Outsourcing. *Information Management and Computer Security, 2*(4), 3–8. doi: 10.1108/09685229410068235. Emerald.

Fomin, V. V., Vries, H., & Barlette, Y. (2008). ISO/IEC 27001 information systems security management standard: exploring the reasons for low adoption. In: *EUROMOT 2008 Conference, Nice, France.*

Fomin, V. V., Vries, H., & Barlette, Y. (2008). ISO/IEC 27001 information systems security management standard: exploring the reasons for low adoption. In: *Proceedings of The Third European Conference on Management of Technology (EUROMOT).*

Fordyce, S. (1982). Computer security: A current assessment. *Computers & Security, 1*(1), 9–16.

Furnell, S. (2005). Why users cannot use security. *Computers & Security, 24*(4), 274–279.

Furnell, S. M., & Karweni, T. (1999). Security implications of electronic commerce: a survey of consumers and businesses. *Internet Research, 9*(5), 372–382.

Furnell, S. M., Jusoh, A., & Katsabas, D. (2006). The challenges of understanding and using security: A survey of end-users. *Computers & Security, 25*(1), 27–35.

Furnell, S. M., Onions, P. D., Knahl, M., Sanders, P. W., Bleimann, U., Gojny, U., & Röder, H. F. (1998). A security framework for online distance learning and training. *Internet Research, 8*(3), 236–242.

Galvan, J. L. (2006). *A Guide for Students of the Social and Behavioral Sciences.* Pyrczak Publishing.

Gan, X. (2006). Software Performance Testing. In: *Seminar Paper, University of Helsinki.* pp. 26–29.

Gillies, A. (2011). Improving the quality of information security management systems with ISO27000. *The TQM Journal, 23*(4), 367–376.

Gollmann, D. (2010). Computer security. *Wiley Interdisciplinary Reviews: Computational Statistics, 2*(5), 544–554.

Gordon, L. A., & Loeb, M. P. (2002). The economics of information security investment. *ACM Transactions on Information and System Security (TISSEC), 5*(4), 438–457.

Green, B. F., & Hall, J. A. (1984). Quantitative methods for literature reviews. *Annual Review of Psychology, 35*(1), 37–54.

Halderman, J. A., Schoen, S. D., Heninger, N., Clarkson, W., Paul, W., & Calandrino, J. A. (2009). Lest we remember: cold-boot attacks on encryption keys. *Communications of the ACM, 52*(5), 91–98.

Hart, T. B. (2008). Visual Methodologies: Rose, Gillian. Visual Methodologies: An Introduction to the Interpretation of Visual Materials. London: Sage, 2007, 301 pp.

Hentea, M. (2005). A perspective on achieving information security awareness. *Informing Science: International Journal of an Emerging Transdiscipline, 2,* 169–178.

Herbsleb, J., Zubrow, D., Goldenson, D., Hayes, W., & Paulk, M. (1997). Software quality and the capability maturity model. *Communications of the ACM, 40*(6), 30–40.

Highsmith, J., & Cockburn, A. (2001). Agile software development: The business of innovation. *Computer, 34*(9), 120–127.

Hoo, K. J. S. (2000). *How Much is Enough? A Risk Management Approach to Computer Security.* Stanford University.

Humphreys, E. J., Moses, R. H., & Plate, A. E. (1998). *Guide to Risk Assessment and Risk Management.* British Standards Institution.

Huo, M., Verner, J., Zhu, L., & Babar, M. A. (2004, September). Software quality and agile methods. In: *Computer Software and Applications Conference, 2004. COMPSAC 2004. Proceedings of the 28th Annual International.* IEEE. pp. 520–525.

Information Assurance Advisory Council (IAAC) in association with Microsoft. Benchmarking Information Assurance. (2002).

Information Assurance Collaboration Group (IACG). (2007). Industry Response To The HMG Information Assurance Strategy and Delivery Plan. A report by the IACG Working Group On The Role Of Industry In Delivering The National IA Strategy (IWI009).

International Standard Organization (ISO). 2004. Information Security Management System. Obtained from http://www.isms-guide.blogspot.com/2007/11/key-components-ofstandard-iso-27001-iso.html and www.ISO 27001security.com.

ISACA. (2008). Glossary of Terms. Available online at http://www.isaca.org/Knowledge-Center/Documents/Glossary/glossary.pdf [accessed on 10.07.2011].

ISACA. (2009). *An Introduction to the Business Model for Information Security.*

ISO. (2004). ISO/IEC 13335–1: Information technology – Security techniques – Management of information and communications technology security. Part 1: Concepts and models for information and communications technology security management.

ISO. (2009). (E) ISO/IEC 27000: Information technology – Security techniques – Information security management systems – Overview and vocabulary.

ISO. (2009). ISO/IEC 15408–1: Information technology – Security techniques – Evaluation criteria for IT security. Part 1: Introduction and general model.

ISO/IEC 27000:2009 (E) Information technology - Security techniques – Information security management systems - Overview and vocabulary.

IT Governance Institute (ITGI). (2007). COBIT 4.1 Excerpts. Rolling Meadows, IL 60008 USA.

IT Governance Institute (ITGI). (2007). COBIT 4.1 Executive Overview. ITGI Rolling Meadows, USA.

IT Governance Institute (ITGI). (2008). Aligning CobiT® 4. 1, ITIL® V3 and ISO/IEC 27002 for Business Benefit A Management Briefing From ITGI and OGC. ITGI Rolling Meadows, USA.

IT Governance Institute (ITGI). (2008). COBIT Mapping. ITGI Rolling Meadows, USA.

IT Governance Institute (ITGI). (2008). Mapping of ITIL v3 with COBIT 4.1. Rolling Meadows, IL 60008 USA.

Jackson, M. (2001). *Problem Frames: Analyzing and Structuring Software Development Problems*. Addison-Wesley.

Jung, H. W., Kim, S. G., & Chung, C. S. (2004). Measuring software product quality: A survey of ISO/IEC 9126. *IEEE Software, 21*(5), 88–92.

Kan, S. H. (2002). *Metrics and Models in Software Quality Engineering*. Addison-Wesley Longman Publishing Co., Inc.

Kankanhalli, A., Teo, H. H., Tan, B. C., & Wei, K. K. (2003). An integrative study of information systems security effectiveness. *International Journal of Information Management, 23*(2), 139–154.

Kawakoya, Y., Iwamura, M., & Itoh, M. (2010, October). Memory behavior-based automatic malware unpacking in stealth debugging environment. In: *2010 5th International Conference on Malicious and Unwanted Software (MALWARE)*, IEEE. pp. 39–46.

Kazemi, M., Khajouei, H., & Nasrabadi, H. (2012). Evaluation of information security management system success factors: Case study of Municipal organization. *African Journal of Business Management, 6*(14), 4982–4989.

Kelleher, Z., & Hall, H. (2005). Response to risk Experts and end-user perspectives on email security, and the role of the business information professional in policy development. *Business Information Review, 22*(1), 46–52.

Khosroshahy, M., Mehmet Ali, M. K., & Qiu, D. (2013). The SIC botnet lifecycle model: A step beyond traditional epidemiological models. *Computer Networks, 57*(2), 404–421.

Kompas.com. (2013). Diplomatic Tensions Indonesia – Australia in Wiretapped issues. Obtained from: www.kompas.com.

Kosutic, D. (2010). ISO 27001 and BS 25999. Obtained from http://blog.ISO 27001standard.com.

Kosutic, D. (2013). *Risk Assessment of ISO 27001*. Retrieved November, 2013, from http://blog.ISO 27001standard.com/.

Kothari, C. R. (2004). *Research Methodology: Methods and Techniques*. New Age International.

Kotler, P. (2002). Marketing Management. Prentice Hall.

Kotler, P., & Levy, S. J. (1969). Broadening the concept of marketing. *Journal of Marketing, 33*(1).

Kruse II, W. G., & Heiser, J. G. (2001). *Computer Forensics: Incident Response Essentials*. Pearson Education.

Lacey, D. (2010). Understanding and transforming organizational security culture. *Information Management & Computer Security*, *18*(1), 4–13.

Lacey, D. (2011). *Managing the Human Factor in Information Security: How to Win Over Staff and Influence Business Managers*. John Wiley & Sons.

Lambo, T. (2006). ISO/IEC 27001: The future of infosec certification. *ISSA Journal, Information Systems Security Organization (http://www.issa.org)*.

Laredo, V. G. (2008). PCI DSS compliance: A matter of strategy. *Card Technology Today*, *20*(4), 9.

Leder, F., Steinbock, B., & Martini, P. (2009). Classification and detection of metamorphic malware using value set analysis. In: *2009 4th International Conference on Malicious and Unwanted Software (MALWARE)*, IEEE. pp. 39–46.

Lee, Y. W., Strong, D. M., Kahn, B. K., & Wang, R. Y. (2002). AIMQ: A methodology for information quality assessment. *Information & Management*, *40*(2), 133–146.

Lichtenstein, S., & Williamson, K. (2006). Understanding consumer adoption of internet banking: An interpretive study in the Australian banking context. *Journal of Electronic Commerce Research*, *7*(2), 50–66.

MacCormack, A., & Verganti, R. (2003). Managing the sources of uncertainty: Matching process and context in software development. *Journal of Product Innovation Management*, *20*(3), 217–232.

Madu, C. N., & Kuei, C. H. (1993). Introducing strategic quality management. *Long Range Planning*, *26*(6), 121–131.

Mataracioglu, T., & Ozkan, S. (2011). Analysis of the User Acceptance for Implementing ISO/IEC 27001: 2005 in Turkish Public Organizations. *arXiv preprint arXiv:1103.0405*.

McCall, J. A., Richards, P. K., & Walters, G. F. (1977a). *Factors in Software Quality. Volume-I. Concepts and Definitions of Software Quality*. General Electric Co., Sunnyvale, CA.

McCall, J. A., Richards, P. K., & Walters, G. F. (1977b). *Factors in Software Quality. Volume-III. Preliminary Handbook on Software Quality for an Acquisition Manager*. General Electric Co., Sunnyvale CA.

McCumber, J. (1991). Information systems security: A comprehensive model. In: *Proceedings of the 14th National Computer Security Conference*.

Montesino, R., Fenz, S., & Baluja, W. (2012). SIEM-based framework for security controls automation. *Information Management & Computer Security*, *20*(4), 248–263.

Morrison, M., Sweeney, A., & Heffernan, T. (2003). Learning styles of on-campus and off-campus marketing students: The challenge for marketing educators. *Journal of Marketing Education*, *25*(3), 208–217.

Morse, E. A., & Raval, V. (2008). PCI DSS: Payment card industry data security standards in context. *Computer Law & Security Review*, *24*(6), 540–554.

Morse, E. A., & Raval, V. (2011). Private ordering in light of the law: Achieving consumer protection through payment card security measures. *DePaul Bus. & Comm. LJ*, *10*, 213.

Müller, D., Herbst, J., Hammori, M., & Reichert, M. (2006). *IT Support for Release Management Processes in the Automotive Industry*. Springer Berlin Heidelberg. pp. 368–377.

Neumann, P. G. (1999). *Practical Architectures for Survivable Systems and Networks: Phase-One Final Report.* Sri International Menlo Park, CA, Computer Science Lab.

Nicolett, M., & Kavanagh, K. M. (2011). Magic quadrant for security information and event management. *Gartner RAS Core Research Note (May 2009).*

Ohki, E., Harada, Y., Kawaguchi, S., Shiozaki, T., & Kagaua, T. (2009). Information security governance framework. *Proceedings of the First ACM Workshop on Information Security Governance.*

Oliver, D., & Lainhart, J. (2012). COBIT 5: Adding Value Through Effective Geit. *EDPACS, 46*(3), 1–12.

Onwubiko, C. (2009). A security audit framework for security management in the enterprise. In: *Global Security, Safety, and Sustainability.* Springer Berlin Heidelberg. pp. 9–17.

Othman, M. F. I., Chan, T., Foo, E., Nelson, K. J., & Timbrell, G. T. (2011, August). Barriers to information technology governance adoption: a preliminary empirical investigation. In: *Proceedings of 15th International Business Information Management Association Conference.* pp. 1771–1787.

Parker, D. B. (1998). *Fighting Computer Crime: A New Framework for Protecting Information* (pp. I–XV). New York: Wiley.

Patton, M. Q. (1980). *Qualitative Evaluation Methods.*

Patton, M. Q. (2005). *Qualitative Research.* John Wiley & Sons, Ltd.

Peltier, T. R. (2005a). *Information Security Risk Analysis.* CRC press.

Peltier, T. R. (2005b). Implementing an Information Security Awareness Program. *Information Systems Security, 14*(2), 37–49.

Peters, J. F., & Pedrycz, W. (1998). *Software Engineering: An Engineering Approach.* John Wiley & Sons, Inc.

Pipkin, D. L. (2000). *Information Security: Protecting the Global Enterprise.* Upper Saddle River, NJ: Prentice Hall PTR.

Pipkin, D. L. (2000). *Information Security: Protecting the Global Enterprise.* Upper Saddle River, NJ: Prentice Hall PTR.

Pollitt, D. (2005). Energies trains employees and customers in IT security: only one company in ten has staff with the necessary qualifications. *Human Resource Management International Digest, 13*(2), 25–28.

Posthumus, S., & Von Solms, R. (2004). A framework for the governance of information security. *Computers & Security, 23*(8), 638–646.

Potter, C., & Beard, A. (2010). Information security breaches survey 2010. *Price Water House Coopers. Earl's Court, London.*

Potter, C., & Beard, A. (2012). Information security breaches survey 2012. *Price Water House Coopers.* Earl's Court, London.

Prince Muqrin Chair (PMC). (2010). Area and Trends of Computer Security Issues.

Puhakainen, P., & Ahonen, R. (2006). Design theory for information security awareness.

Queensland Government Information Security Policy Framework (QGISPF). (2009). *Shaping Government ICT to Support Business Outcomes: Queensland Government Information Security Policy Framework.* Queensland Government.

Rahim, A., & Bin Muhaya, F. T. (2010). Discovering the Botnet Detection Techniques. In *Security Technology, Disaster Recovery and Business Continuity.* Springer Berlin Heidelberg. pp. 231–235.

Ralph, P., & Wand, Y. (2009). A proposal for a formal definition of the design concept. In: *Design Requirements Engineering: A Ten-Year Perspective.* Springer Berlin Heidelberg. pp. 103–136.

Ramayah, T., Jantan, M., Mohd Noor, M. N., Razak, R. C., & Koay, P. L. (2003). Receptiveness of internet banking by Malaysian consumers: The case of Penang. *Asian Academy of Management Journal, 8*(2), 1–29.

Ridley, G., Young, J., & Carroll, P. (2004, January). COBIT and its Utilization: A framework from the literature. In: *Proceedings of the 37th Annual Hawaii International Conference on System Sciences, 2004.* IEEE. 8 pp.

Rowlingson, R., & Winsborrow, R. (2006). A comparison of the Payment Card Industry data security standard with ISO17799. *Computer Fraud & Security, 3,* 16–19.

Saint-Germain, R. (2005). Information security management best practice based on ISO/IEC 17799. *Information Management Journal, 39*(4), 60–66.

Saleh, M. S., & Alfantookh, A. (2011). A new comprehensive framework for enterprise information security risk management. *Applied Computing and Informatics, 9*(2), 107–118.

Saleh, M. S., Alrabiah, A., & Bakry, S. H. (2007a). A STOPE model for the investigation of compliance with ISO 17799–2005. *Information Management & Computer Security, 15*(4), 283–294.

Saleh, M. S., Alrabiah, A., & Bakry, S. H. (2007b). Using ISO 17799: 2005 information security management: a STOPE view with six sigma approach. *International Journal of Network Management, 17*(1), 85–97.

Scandura, T. A., & Williams, E. A. (2000). Research methodology in management: Current practices, trends, and implications for future research. *Academy of Management Journal, 43*(6), 1248–1264.

Schneier, B. (1999). Attack trees. *Dr. Dobb's Journal, 24*(12), 21–29.

Schneier, B. (2001). Secrets & Lies: Digital Security in a Networked World. *International Hydrographic Review, 2*(1), 103–104.

Schneier, B. (2008). The psychology of security. In: *Progress in Cryptology–AFRICAC-RYPT 2008.* Springer Berlin Heidelberg. pp. 50–79.

Schneier, B. (2009). *Schneier on Security.* John Wiley & Sons.

Schwalbe, K. (2010). *Information Technology Project Management, Revised.* Cengage Learning.

Schweitzer, J. A. (1982). *Managing Information Security: A Program for the Electronic Information Age.* Butterworth Publishers.

Shahsavarani, N., & Ji, S. (2014). Research in Information Technology Service Management (ITSM) (2000–2010): An Overview. *International Journal of Information Systems in the Service Sector (IJISSS), 6*(4), 73–91.

Shaw, A. (2009). Data breach: From notification to prevention using PCI DSS. *Colum. JL & Soc. Probs., 43,* 517.

Shaw, M. L. N. M. J., & Strader, T. J. (2010). Sustainable e-Business Management.

Sherwood, J., Clark, A., & Lynas, D. (2005). Enterprise security architecture. *Computer Security Journal, 21*(4), 24.

Shieh, S. P., & Gligor, V. D. (1997). On a pattern-oriented model for intrusion detection. *IEEE Transactions on Knowledge and Data Engineering, 9*(4), 661–667.

Shoemaker, D., Bawol, J., Drommi, A., & Schymik, G. (2004). A delivery model for an Information Security curriculum. In: *Proceedings of the Third Security Conference.*

Sipior, J. C., & Ward, B. T. (2008). *A Framework for Information Security Management Based on Guiding Standards: A United States Perspective*. Issues in Informing Science & Information Technology, p. 5.

Siponen, M. (2006). Six design theories for IS security policies and guidelines. *Journal of the Association for Information Systems*, *7*(1), 19.

Siponen, M., & Willison, R. (2007). A critical assessment of IS security research between 1990–2004.

Siponen, M., & Willison, R. (2009). Information security management standards: Problems and solutions. *Information & Management*, *46*(5), 267–270.

Standish Group. (2013). ICT Project Report. The CHAOS Manifesto.

Straub, D. W., & Welke, R. J. (1998). Coping with systems risk: Security planning models for management decision making. *Mis Quarterly*, 441–469.

Straub, D., Keil, M., & Brenner, W. (1997). Testing the technology acceptance model across cultures: A three country study. *Information & Management*, *33*(1), 1–11.

Straub, Jr., D. W. (1990). Effective IS security: An empirical study. *Information Systems Research*, *1*(3), 255–276.

Susanto, H., & Almunawar, M. N., (2012d). Information Security Awareness Within Business Environment: An IT Review. ASEAN FBEPS-AGBEP PhD Colloquium. UBD, Brunei Darussalam.

Susanto, H., & Muhaya, F. B. (2010a). Multimedia Information Security Architecture Framework. In: *2010 5th International Conference on Future Information Technology (FutureTech),* IEEE. pp. 1–6.

Susanto, H., Almunawar, M. N., & Kang, C. C. (2012e). Toward Cloud Computing Evolution: Efficiency vs. Trendy vs. Security. *International Journal of Engineering and Technology – UK. 2*(9). Available at SSRN 2039739.

Susanto, H., Almunawar, M. N., & Tuan, Y. C. (2011a). Information Security Management System Standards: A Comparative Study of the Big Five. *International Journal of Electrical & Computer Sciences, 11*(5).

Susanto, H., Almunawar, M. N., & Tuan, Y. C. (2011b). I-SolFramework View on ISO 27001. Information Security Management System: Refinement Integrated Solution's Six Domains. *Journal of Computer, Asian Transaction.*

Susanto, H., Almunawar, M. N., & Tuan, Y. C. (2012b). Information security challenge and breaches: novelty approach on measuring ISO 27001 readiness level. *International Journal of Engineering and Technology*, *2*(1).

Susanto, H., Almunawar, M. N., & Tuan, Y. C. (2012c). A novel method on ISO 27001 reviews: ISMS compliance readiness level measurement. *arXiv preprint arXiv:1203.6622.*

Susanto, H., Almunawar, M. N., & Tuan, Y. C. (2012f). Information security challenge and breaches: novelty approach on measuring ISO 27001 readiness level. *International Journal of Engineering and Technology*, *2*(1). *Preprint arXiv:1203.6622.*

Susanto, H., Almunawar, M. N., & Tuan, Y. C. (2012h). Information security challenge and breaches: novelty approach on measuring ISO 27001 readiness level. *International Journal of Engineering and Technology*, *2*(1).

Susanto, H., Almunawar, M. N., Tuan, Y. C., & Aksoy, M. S. (2012g). I-SolFramework: An Integrated Solution Framework Six Layers Assessment on Multimedia Information Security Architecture Policy Compliance. *International Journal of Electrical & Computer Sciences, 12*(1).

Susanto, H., Almunawar, M. N., Tuan, Y. C., Aksoy, M. S., & Syam, W. P. (2012a). Integrated solution modeling software: A new paradigm on information security review and assessment. *arXiv preprint arXiv:1203.6214*.

Susanto, H., Muhaya,F., & Almunawar, M. N. (2010b). Refinement of Strategy and Technology Domains STOPE View on ISO 27001. Accepted paper, International Conference on Intelligent Computing and Control – Future Technology (ICOICC 2010). Archived *preprint arXiv:1204.1385*.

The European Union Agency for Network and Information Security (ENISA). Information Security Awareness. Obtained from www.enisa.europa.eu. November 2012.

Theoharidou, M., Kokolakis, S., Karyda, M., & Kiountouzis, E. (2005). The insider threat to information systems and the effectiveness of ISO17799. *Computers & Security, 24*(6), 472–484.

Thomson, M. E., & von Solms, R. (1998). Information security awareness: Educating your users effectively. *Information Management & Computer Security, 6*(4), 167–173.

Tiller, J. S. (2010). *Adaptive Security Management Architecture*. CRC Press.

Toleman, M., Cater-Steel, A., Kissell, B., Chown, R., & Thompson, M. (2009). Improving ICT governance: A radical restructure using CobiT and ITIL. *Information Technology Governance and Service Management: Frameworks and Adaptations, Information Science Reference, Hershey*, 178–189.

Trend Micro. (2011). In: *Internet Content Security Software and Cloud Computing Security.* Obtained from: www.trendmicro.com

TribunNews.com. (2013). Indonesian ICT Markets. Obtained from www.tribunnews.com.

Tsiakis, T., & Stephanides, G. (2005). The economic approach of information security. *Computers & Security, 24*(2), 105–108.

Ungoed-Thomas, J. (2003). The e-mail timebomb. *Sunday Times*, p. 19.

Van Vliet, H., Van Vliet, H., & Van Vliet, J. C. (1993). *Software Engineering: Principles and Practice* (Vol. 3). Wiley.

Von Solms, B. (2001). Information security–a multidimensional discipline. *Computers & Security, 20*(6), 504–508.

Von Solms, B. (2005). *Information Security Governance: COBIT or ISO 17799 or Both?.* Computers & Security, 24(2), 99–104. Elsevier.

Von Solms, B. (2005b). Information Security governance: COBIT or ISO 17799 or both?. *Computers & Security, 24*(2), 99–104.

Von Solms, B., & von Solms, R. (2005). From information security to business security?. *Computers & Security, 24*(4), 271–273.

Von Solms, S. H. (2005a). Information security governance–compliance management vs. operational management. *Computers & Security, 24*(6), 443–447.

Wahono, R. S. (2006). *Software Quality Measurement Techniques*. Obtained from: www.ilmu-komputer.com

Walenstein, A., Hefner, D. J., & Wichers, J. (2010, October). Header information in malware families and impact on automated classifiers. In: *2010 5th International Conference on Malicious and Unwanted Software (MALWARE)*, IEEE. pp. 15–22

Whitman, M., & Mattord, H. (2011). *Principles of Information Security*. Cengage Learning.

Whitten, J. L., Barlow, V. M., & Bentley, L. (1997). *Systems Analysis and Design Methods.* McGraw-Hill Professional.

Wolfgang, P. (1994). *Design Patterns for Object-Oriented Software Development.* Reading, Mass: Addison-Wesley.

Woon, I. M., & Kankanhalli, A. (2007). Investigation of IS professionals' intention to practice secure development of applications. *International Journal of Human-Computer Studies,* *65*(1), 29–41.

Wu, D. D., & Olson, D. L. (2009). Introduction to the special section on "optimizing risk management: methods and tools". *Human and Ecological Risk Assessment, 15*(2), 220–226.

Yasinsac, A., Erbacher, R. F., Marks, D. G., Pollitt, M. M., & Sommer, P. M. (2003). Computer forensics education. *IEEE Security & Privacy Magazine, 1*(4), 15–23.

Zeltser, L., Skoudis, E., Stratton, W., & Teall, H. (2003). Malware: Fighting Malicious Code.

# INDEX

9 781774 636527